SKILLS MASTERY

This Book Includes:

- Practice questions to help students master topics assessed on the the PARCC and Smarter Balanced Tests
  - Operations and Algebraic Thinking
  - Numbers and Operations in Base Ten
  - Numbers and Operations – Fractions
  - Measurement and Data
  - Geometry
- Detailed answer explanations for every question
- Strategies for building speed and accuracy
- Content aligned with the Common Core State Standards

Plus access to Online Workbooks which include:

- Hundreds of practice questions
- Self-paced learning and personalized score reports
- Instant feedback after completion of the workbook

Complement Classroom Learning All Year

Using the Lumos Study Program, parents and teachers can reinforce the classroom learning experience for children. It creates a collaborative learning platform for students, teachers, and parents.

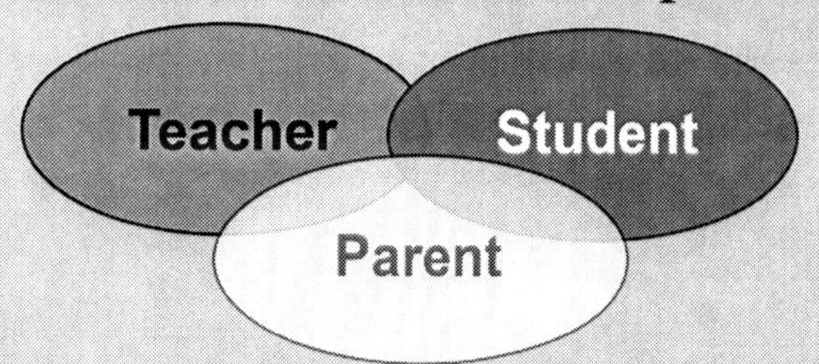

Used in Schools and Libraries
To Improve Student Achievement

Lumos Learning

Common Core Practice - Grade 5 Math: Workbooks to Prepare for the PARCC or Smarter Balanced Test

Contributing Author - April LoTempio
Curriculum Director - Marisa Adams
Executive Producer - Mukunda Krishnaswamy
Designer - Mirona Jova
Database Administrator - R. Raghavendra Rao

ISBN-10: 1940484456

ISBN-13: 978-1-940484-45-7

Printed in the United States of America

For permissions and additional information contact us

Lumos Information Services, LLC
PO Box 1575, Piscataway, NJ 08855-1575
http://www.LumosLearning.com

Email: support@lumoslearning.com
Tel: (732) 384-0146
Fax: (866) 283-6471

Lumos Learning 

Table of Contents

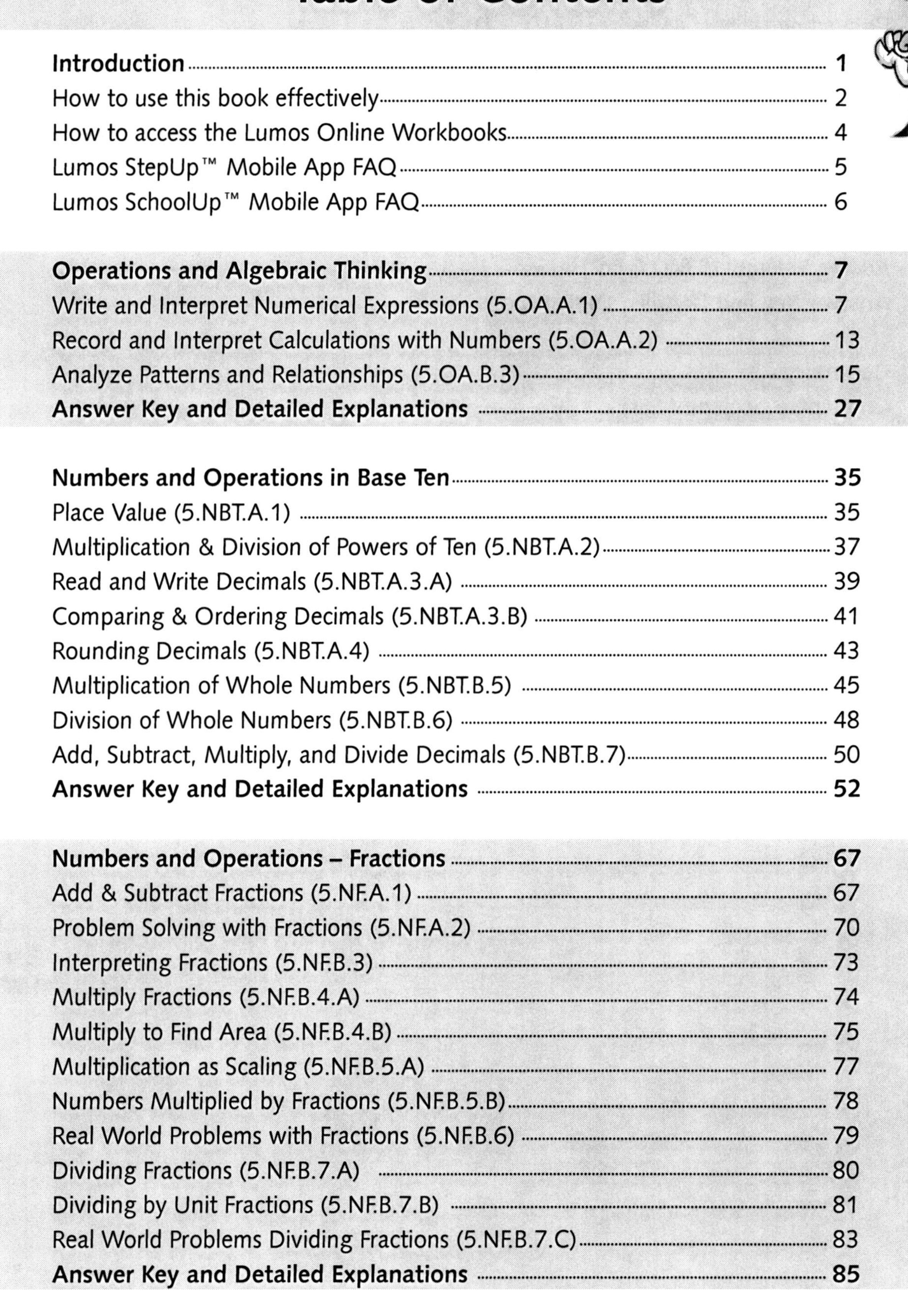

Introduction

The Common Core State Standards Initiative (CCSS) was created from the need to have more robust and rigorous guidelines, which could be standardized from state to state. These guidelines create a learning environment where students will be able to graduate high school with all skills necessary to be active and successful members of society, whether they take a role in the workforce or in some sort of post-secondary education.

Once the CCSS were fully developed and implemented, it became necessary to devise a way to ensure they were assessed appropriately. To this end, states adopting the CCSS have joined one of two consortia, either PARCC or Smarter Balanced.

Why Practice by Standard?

Each standard, and substandard, in the CCSS has its own specific content. Taking the time to study and practice each one individually can help students more adequately understand the CCSS for their particular grade level. Additionally, students have individual strengths and weaknesses. Being able to practice content by standard allows them the ability to more deeply understand each standard and be able to work to strengthen academic weaknesses.

How Can the Lumos Study Program Prepare Students for Standardized Tests?

Since the fall of 2014, student mastery of Common Core State Standards are being assessed using standardized testing methods. At Lumos Learning, we believe that yearlong learning and adequate practice before the actual test are the keys to success on these standardized tests. We have designed the Lumos study program to help students get plenty of realistic practice before the test and to promote yearlong collaborative learning.

This is a Lumos tedBook™. It connects you to Online Workbooks and additional resources using a number of devices including android phones, iPhones, tablets and personal computers. Each Online Workbook will have some of the same questions seen in this printed book, along with additional questions. The Lumos StepUp® Online Workbooks are designed to promote yearlong learning. It is a simple program students can securely access using a computer or device with internet access. It consists of hundreds of grade appropriate questions, aligned to the new Common Core State Standards. Students will get instant feedback and can review their answers anytime. Each student's answers and progress can be reviewed by parents and educators to reinforce the learning experience.

How to use this book effectively

The Lumos Program is a flexible learning tool. It can be adapted to suit a student's skill level and the time available to practice before standardized tests. Here are some tips to help you use this book and the online resources effectively:

Students

- The standards in each book can be practiced in the order designed, or in the order of your own choosing.
- Complete all problems in each workbook.
- Use the Online workbooks to further practice your areas of difficulty and complement classroom learning.
- Download the Lumos StepUp® app using the instructions provided to have anywhere access to online resources.
- Practice full length tests as you get closer to the test date.
- Complete the test in a quiet place, following the test guidelines. Practice tests provide you an opportunity to improve your test taking skills and to review topics included in the CCSS related standardized test.

Parents

- Familiarize yourself with your state's consortium and testing expectations.
- Get useful information about your school by downloading the Lumos SchoolUp™ app. Please follow directions provided in "How to download Lumos SchoolUp™ App" section of this chapter.
- Help your child use Lumos StepUp® Online Workbooks by following the instructions in "How to access the Lumos Online Workbooks" section of this chapter.
- Help your student download the Lumos StepUp® app using the instructions provided in "How to download the Lumos StepUp® Mobile App" section of this chapter.
- Review your child's performance in the Lumos Online Workbooks periodically. You can do this by simply asking your child to log into the system online and selecting the subject area you wish to review.
- Review your child's work in each workbook.

Teachers

- You can use the Lumos online programs along with this book to complement and extend your classroom instruction.

- Get a Free Teacher account by using the respective states specific links and QR codes below:

PARCC States	SBAC States
LumosLearning.com/a/stepupbasic	**LumosLearning.com/a/sbacbasic**

This Lumos StepUp® Basic account will help you:

- Create up to 30 student accounts.
- Review the online work of your students.
- Easily access CCSS.
- Create and share information about your classroom or school events.

NOTE: There is a limit of one grade and subject per teacher for the free account.

- Download the Lumos SchoolUp™ mobile app using the instructions provided in "How can I Download the App?" section of this chapter.

How to Access the Lumos Online Workbooks

First Time Access:

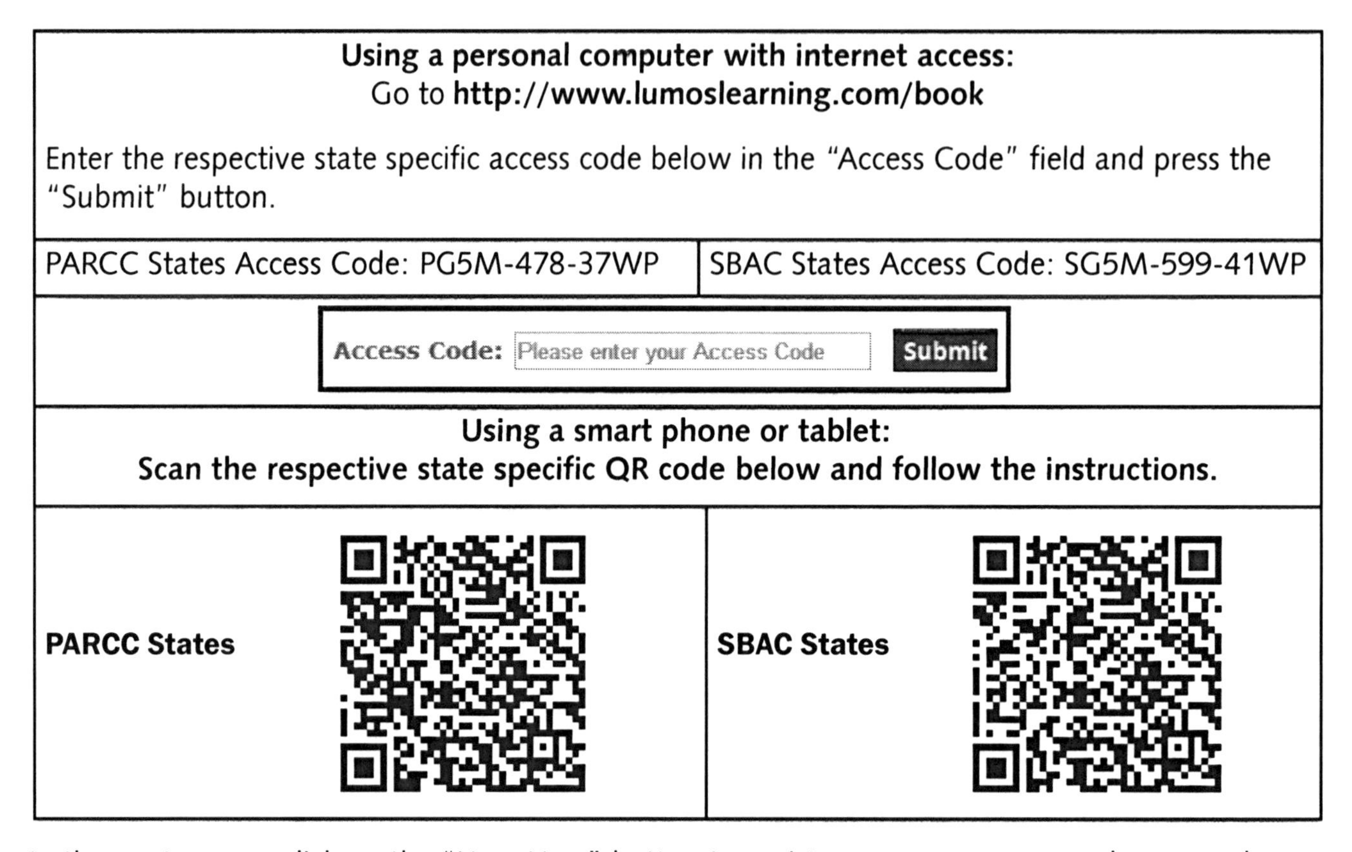

Using a personal computer with internet access:
Go to **http://www.lumoslearning.com/book**

Enter the respective state specific access code below in the "Access Code" field and press the "Submit" button.

PARCC States Access Code: PG5M-478-37WP	SBAC States Access Code: SG5M-599-41WP

Using a smart phone or tablet:
Scan the respective state specific QR code below and follow the instructions.

PARCC States	**SBAC States**

In the next screen, click on the "New User" button to register your user name and password.

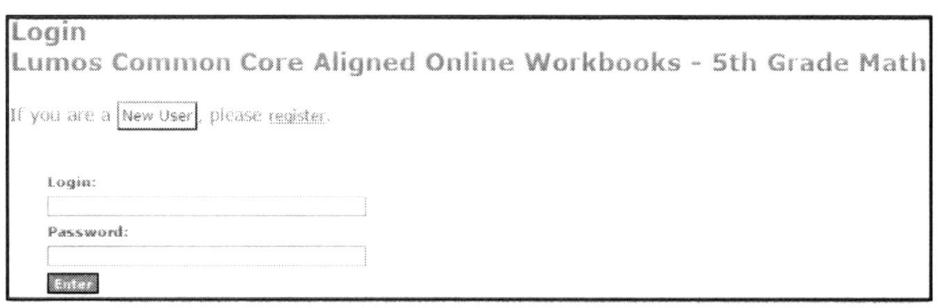

Subsequent Access:

After you establish your user id and password for subsequent access, simply login with your account information.

What if I buy more than one Lumos Study Program?

Please note that you can use all Online Workbooks with one User ID and Password. If you buy more than one book, you will access them with the same account.

Go back to the **http://www.lumoslearning.com/book** link and enter the access code provided in the second book. In the next screen simply login using your previously created account.

Lumos StepUp® Mobile App FAQ For Students

What is the Lumos StepUp® App?

It is a FREE application you can download onto your Android smart phones, tablets, iPhones, and iPads.

What are the Benefits of the StepUp® App?

This mobile application gives convenient access to Practice Tests, Common Core State Standards, Online Workbooks, and learning resources through your smart phone and tablet computers.

- Eleven Technology enhanced question types in both MATH and ELA
- Sample questions for Arithmetic drills
- Standard specific sample questions
- Instant access to the Common Core State Standards
- Jokes and cartoons to make learning fun!

Do I Need the StepUp® App to Access Online Workbooks?

No, you can access Lumos StepUp® Online Workbooks through a personal computer. The StepUp® app simply enhances your learning experience and allows you to conveniently access StepUp® Online Workbooks and additional resources through your smart phone or tablet.

How can I Download the App?

Visit **lumoslearning.com/a/stepup-app** using your smart phone or tablet and follow the instructions to download the app.

QR Code for Smart Phone Or Tablet Users

Lumos SchoolUp™ Mobile App FAQ For Parents and Teachers

What is the Lumos SchoolUp™ App?

It is a FREE App that helps parents and teachers get a wide range of useful information about their school. It can be downloaded onto smartphones and tablets from popular App Stores.

What are the Benefits of the Lumos SchoolUp™ App?

It provides convenient access to

- School "Stickies". A Sticky could be information about an upcoming test, homework, extra curricular activities and other school events. Parents and educators can easily create their own sticky and share with the school community.
- Common Core State Standards.
- Educational blogs.
- StepUp™ student activity reports.

How can I Download the App?

Visit **lumoslearning.com/a/schoolup-app** using your smartphone or tablet and follow the instructions provided to download the App. Alternatively, scan the QR Code provided below using your smartphone or tablet computer.

QR Code
for Smart Phone
Or Tablet Users

Name: ____________________ Date: ____________

Operations and Algebraic Thinking

Write and Interpret Numerical Expressions (5.OA.A.1)

1. $5 + x = 13$
   What is the value of x in this equation?

   Ⓐ $x = 13$
   Ⓑ $x = 18$
   Ⓒ $x = 7$
   Ⓓ $x = 8$

2. What is the word form of 15.045?

   Ⓐ Fifteen and four hundredths and five thousandths
   Ⓑ Fifteen and forty-five hundredths
   Ⓒ Fifteen and forty-five thousandths
   Ⓓ Fifteen and zero forty-five

3. Which of the following situations would be expressed using a negative number?

   Ⓐ A drop in temperature below zero
   Ⓑ A quarterback being tackled for a 5-yard loss
   Ⓒ A treasure chest located 200 ft. below sea level
   Ⓓ All of the above

4. Identify the equation that could be used to solve this problem.

   A set of 72 CDs is being shared equally among 9 people. How many CDs will each person get?

   Ⓐ $72 \div 9 = n$
   Ⓑ $n = 72 + 9$
   Ⓒ $n - 72 = 9$
   Ⓓ $72 \times 9 = n$

5. **Which number sentence could be used to find out how many teddy bears Julie's aunt gave her?**
Julie had a collection of 25 teddy bears. Her aunt gave her some more teddy bears. Now she has 32 teddy bears.

Ⓐ **m - 25 = 32**
Ⓑ **25 + m = 32**
Ⓒ **25 - m = 32**
Ⓓ **m = 25 + 32**

6. **If 7 + x = 21, then what is the value of x?**

Ⓐ **x = 14**
Ⓑ **x = 15**
Ⓒ **x = 28**
Ⓓ **x = 13**

7. **Use the table below to answer the following question.**
If x = 10, what is the value of y?

x	y
0	-2
3	7
4	10
9	25
12	34
10	
5	
	16

Ⓐ **y = 32**
Ⓑ **y = 25**
Ⓒ **y = 28**
Ⓓ **y = 29**

8. **What is the standard form of this number?**
Seventy-nine million, four hundred seventeen thousand, six hundred eight

Ⓐ **79,471,608**
Ⓑ **79,417,068**
Ⓒ **7,941,768**
Ⓓ **79,417,608**

9. **Which of the following number sentences models the Associative Property of Multiplication?**

Ⓐ 80 x 5 = (40 x 5) + (40 x 5)
Ⓑ (11 x 6) x 7 = 11 x (6 x 7)
Ⓒ 3 x 4 x 2 = 2 x 4 x 3
Ⓓ 44 x 1 = 44

10. **Which value(s) of x make this inequality true?**
405 - x > 300

Ⓐ 100
Ⓑ 105
Ⓒ 110
Ⓓ All of the above

11. **Which values of m and n make this inequality true?**
mn ≥ 36

Ⓐ m = 10, n = 4
Ⓑ m = 12, n = 3
Ⓒ m = 1, n = 36
Ⓓ All of the above

12. **What is the value of p in the following equation?**
p - 85 = 90

Ⓐ p = 185
Ⓑ p = 175
Ⓒ p = 5
Ⓓ p = 15

13. **15 ÷ a = 5**
The value of a is ________ .

Ⓐ a = 75
Ⓑ a = 3
Ⓒ a = 15
Ⓓ a = 5

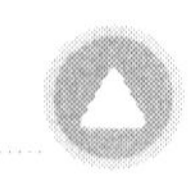

14. Which equation could be used to solve this problem?

On Friday, 120 people shopped at the Sports Shack. This is twice the number of people who had shopped there on Thursday. How many people shopped at the Sports Shack on Thursday?

Ⓐ **120 = n x 2**
Ⓑ **120 x 2 = n**
Ⓒ **120 = n + 2**
Ⓓ **n x 120 = 2**

15. Write an equation that could be used to solve this problem.

Paul had 52 crayons. He gave some of his crayons to his sister. Now he has 38 crayons left. How many crayons did he give to his sister?

Ⓐ **38 - c = 52**
Ⓑ **52 + c = 38**
Ⓒ **52 - c = 38**
Ⓓ **38 + 52 = c**

16. The rule for the following table is, "y equals 9 less than the triple of x." When x = 16, what is the value of y?

x	y
0	-9
3	
8	15
11	24
16	

Ⓐ **y = 29**
Ⓑ **y = 41**
Ⓒ **y = 39**
Ⓓ **y = 37**

Name: ____________________ Date: ____________

17. The rule for the following function is: "y is equal to one less than the square of x."
Use the table to respond to the following:
What x value would result in a y value of 80?

x	y
0	-1
1	0
2	3
3	8
4	
7	
	80

Ⓐ 10
Ⓑ 9
Ⓒ 8
Ⓓ 11

18. The pattern below has the rule _____ .
32, 44, 56, 68,

Ⓐ **Add 32**
Ⓑ **Add 6**
Ⓒ **Add 12**
Ⓓ **Add 8**

19. Abbottsville, a medium sized suburban community, has seen a steady increase in its population over the past 15 years. In 1995, the population was around 25,000. The population has increased at a rate of about 1,000 people per year since then. **The graph below shows this trend. Use the graph to answer the following question:**
If the trend continues as shown, what is the expected population of Abbottsville in 2015?

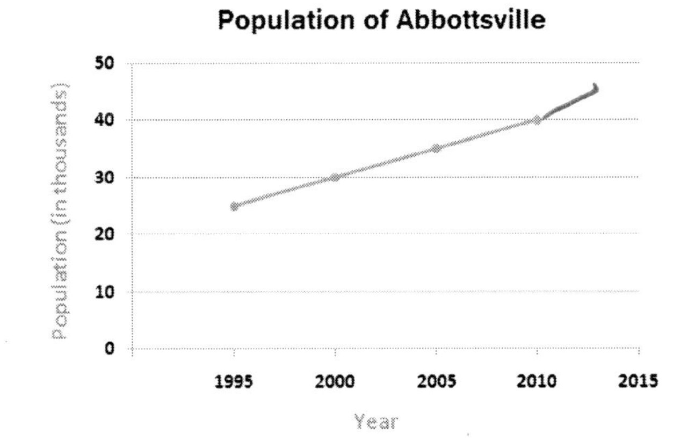

Ⓐ 45,000
Ⓑ 41,000
Ⓒ 40,000
Ⓓ 50,000

20. Kevin has been cutting lawns to earn some extra spending money. The first week he worked, he earned $10.00. Each successive week, for the next three weeks, he earned twice what he had earned the week before. How much money, in all, did he earn during the first four weeks of work?

Ⓐ $70.00
Ⓑ $80.00
Ⓒ $120.00
Ⓓ $150.00

Name: ______________________ Date: ______________

Record and Interpret Calculations with Numbers (5.OA.A.2)

1. Which expression shows 10 more than the "quotient" of 72 "divided" by 8?

   Ⓐ (10 + 72) ÷ 8
   Ⓑ (72 ÷ 8) + 10
   Ⓒ 72 ÷ (8 + 10)
   Ⓓ 8 ÷ (72 + 10)

2. Which expression shows 75 minus the product of 12 and 4?

   Ⓐ (75 – 12) x 4
   Ⓑ (12 x 4) – 75
   Ⓒ 75 – (12 + 4)
   Ⓓ 75 – (12 x 4)

3. Jamie purchased 10 cases of soda for a party. Each case holds 24 cans. He also purchased 3 six packs of juice. Which expression best represents the number of cans he purchased?

   Ⓐ (10 x 24) + (3 x 6)
   Ⓑ (10 + 24) x (3 + 6)
   Ⓒ 10 x (24 + 6)
   Ⓓ 10 x 24 x 3 x 6

4. Olivia had 42 pieces of candy. She kept 9 pieces for herself and then divided the rest evenly among her three friends. Which expression best represents the number of candy each friend received?

   Ⓐ (42 ÷ 3) - 9
   Ⓑ (42 – 9) ÷ 3
   Ⓒ 42 ÷ (9 – 3)
   Ⓓ 42 – (9 ÷ 3)

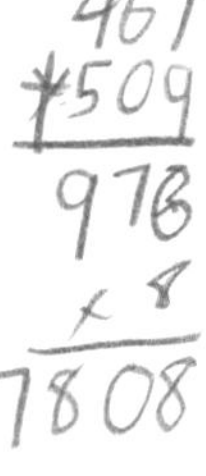

5. Which is true about the solution to 8 x (467 + 509)?

   Ⓐ It is a number in the ten thousands.
   Ⓑ It is an odd number.
   Ⓒ It is eight times greater than the sum of 467 and 509.
   Ⓓ It is 509 more than the product of 8 and 467.

6. **Which is true about the solution to (3,258 – 741) ÷ 3?**

   Ⓐ **It is one third as much as the difference between 3,258 and 741.**
   Ⓑ **It is 741 less than the quotient of 3,258 divided by 3.**
   Ⓒ **It is not a whole number.**
   Ⓓ **It is a number in the thousands.**

7. **Which of these expressions would result in the greatest number?**

   Ⓐ **420 – (28 x 13)**
   Ⓑ **420 + 28 + 13**
   Ⓒ **(420 – 28) x 13**
   Ⓓ **420 + (28 x 13)**

8. **Which of these expressions would result in the smallest number?**

   Ⓐ **684 – (47 + 6)**
   Ⓑ **684 – 47 – 6**
   Ⓒ **(684 – 47) x 6**
   Ⓓ **684 – (47 x 6)**

9. **Each of the 25 students in a class sold 7 items for a fundraiser. Their teacher also sold 13 items. Which expression best represents the number of items they sold in all?**

   Ⓐ **25 x (7 + 13)**
   Ⓑ **13 + (25 x 7)**
   Ⓒ **7 x (25 + 13)**
   Ⓓ **25 + 7 + 13**

10. **Mario had $75. He doubled that amount by mowing his neighbor's lawn all summer. Then he spent $47 on new sneakers. Which expression best represents the amount of money he now has?**

   Ⓐ **(75 x 2) - 47**
   Ⓑ **(75 + 75) ÷ 47**
   Ⓒ **47 – (75 + 2)**
   Ⓓ **75 + 2 - 47**

Analyze Patterns and Relationships (5.OA.B.3)

1. **Which set of numbers completes the function table?**

Rule: x 3

Input	Output
1	☐
2	☐
5	15
8	☐
12	☐

Ⓐ 4, 5, 11, 15
Ⓑ 3, 6, 24, 36
Ⓒ 3, 6, 32, 48
Ⓓ 11, 12, 18, 112

2. **Which set of numbers completes the function table?**

Rule: + 4, ÷ 2

Input	Output
4	☐
6	☐
10	7
22	☐
40	☐

Ⓐ 1, 3, 19, 37
Ⓑ 10, 12, 28, 46
Ⓒ 4, 5, 13, 22
Ⓓ 16, 20, 52, 88

Name: ______________________ Date: ______________

3. **Which set of coordinate pairs matches the function table?**

Rule: x 2, – 1

Input	Output
5	☐
9	17
14	☐
25	☐

Ⓐ (5 , 9), (9 , 17), (14 , 27), (25 , 49)
Ⓑ (5 , 9), (14 , 25), (9 , 17), (27 , 49)
Ⓒ (5 , 9), (9 , 17), (17 , 14), (14 , 25)
Ⓓ (5 , 11), (9 , 17), (14 , 29), (25 , 51)

4. **Which set of coordinate pairs matches the function table?**

Rule: ÷ 3, + 2

Input	Output
9	☐
15	7
27	☐
33	☐

Ⓐ (9 , 1), (15 , 7), (27 , 19), (33 , 25)
Ⓑ (9 , 5), (15 , 7), (27 , 11), (33 , 13)
Ⓒ (9 , 11), (15 , 7), (27 , 29), (33 , 35)
Ⓓ (9 , 15), (15 , 7), (7 , 27), (27 , 33)

5. **Which set of numbers completes the function table?**

Rule: -4

Input	Output
☐	1
7	3
☐	7
☐	10
☐	15

Ⓐ 0, 3, 6, 11
Ⓑ 3, 10, 17, 25
Ⓒ 4, 28, 40, 60
Ⓓ 5, 11, 14, 19

6. **Which set of numbers completes the function table?**

Rule: + 1, x 5

Input	Output
☐	5
2	15
☐	20
☐	35
☐	55

Ⓐ 2, 5, 15, 20
Ⓑ 1, 4, 7, 11
Ⓒ 30, 105, 180, 280
Ⓓ 0, 3, 6, 10

7. **Which rule describes the function table?**

x	y
11	5
14	8
21	15
28	15

Ⓐ **Add 3**
Ⓑ **Subtract 6**
Ⓒ **Subtract 1, Divide by 2**
Ⓓ **Divide by 2, Add 1**

8. **Which rule describes the function table?**

x	y
4	4
7	10
13	22
20	36

Ⓐ **2x, -4**
Ⓑ **+0**
Ⓒ **+3**
Ⓓ **-1, 2x**

9. **Which describes the graph of this function plotted on a coordinate grid?**

x	y
11	5
14	8
21	15
28	22

Ⓐ A curving line
Ⓑ A horizontal line
Ⓒ An upward sloping line
Ⓓ A downward sloping line

10. **Which type of function would result in a graph that looks like this?**

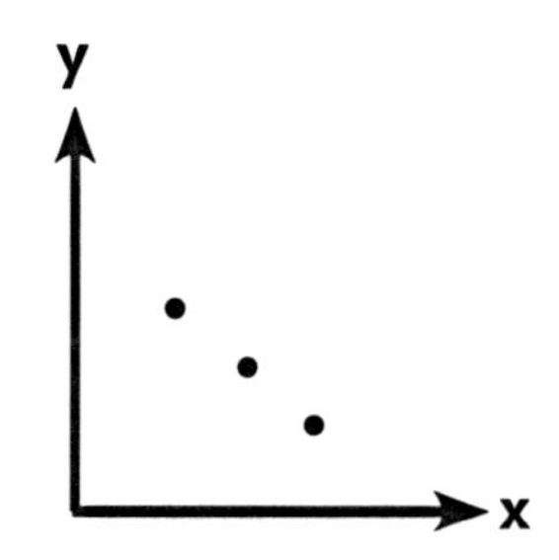

Ⓐ One in which x and y increase at fixed rates
Ⓑ One in which x and y decrease at fixed rates
Ⓒ One in which x decreases while y increases
Ⓓ One in which x increases while y decreases

11. Consider the following two number sequences:
    x: begin at 2, add 3
    y: begin at 4, add 6
    Which describes the relationship between the number sequences?

    Ⓐ The terms in sequence y are two more than the terms in sequence x.
    Ⓑ The terms in sequence y are six times the terms in sequence x.
    Ⓒ The terms in sequence y are two times the terms in sequence x.
    Ⓓ The terms in sequence y are half as much as the terms in sequence x.

12. Consider the following two number sequences:
    x: begin at 1, multiply by 2
    y: begin at 2, multiply by 2
    Which describes the relationship between the number sequences?

    Ⓐ The terms in sequence y are one more than the terms in sequence x.
    Ⓑ The terms in sequence y are two times the terms in sequence x.
    Ⓒ The terms in sequence y are two more than the terms in sequence x.
    Ⓓ The terms in sequence y are half as much as the terms in sequence x.

13. Consider the following number sequence:
    x: begin at 5, add 6
    Which would result in a relationship in which y is always three more than x?

    Ⓐ y: begin at 8, add 6
    Ⓑ y: begin at 5, add 9
    Ⓒ y: begin at 8, add 9
    Ⓓ y: begin at 2, add 6

14. Consider the following number sequence:
    x: begin at 4, multiply by 2
    Which would result in a relationship in which y is always half as much as x?

    Ⓐ y: begin at 4, multiply by 4
    Ⓑ y: begin at 4, multiply by ½
    Ⓒ y: begin at 2, multiply by 1
    Ⓓ y: begin at 2, multiply by 2

15. Which rule corresponds to the following set of coordinate pairs?
    (12 , 7), (14 , 9), (18 , 13)

    Ⓐ Subtract 5
    Ⓑ Add 2
    Ⓒ Multiply by 2
    Ⓓ Add 5

16. Which rule corresponds to the following set of coordinate pairs?
(3 , 9), (5 , 13), (8 , 19)

Ⓐ 3x
Ⓑ +2, +3
Ⓒ 2x, +3
Ⓓ +6

17. Which is true about the following set of coordinate pairs?
(4 , 7), (9 , 12), (11 , 14)

Ⓐ y is always 1/3 as much as x
Ⓑ x is always 3 more than y
Ⓒ x is always 3 less than y
Ⓓ y is always 3 times as much as y

18. Which is true about the following set of coordinate pairs?
(2 , 8), (5 , 20), (10 , 40)

Ⓐ y is always 1/4 as much as x
Ⓑ x is always 6 more than y
Ⓒ x is always 4 more than y
Ⓓ y is always 4 times as much as x

19. The rule is y = 3x + 2. Which number sequence completes the following set of coordinate pairs?
(1 , _), (4 , _), (8 , _)

Ⓐ 33, 36, 40
Ⓑ 5, 14, 26
Ⓒ 1, 10, 22
Ⓓ 6, 9, 13

20. The rule is x = y + 5. Which number sequence completes the following set of coordinate pairs?
(7 , _), (11 , _), (20 , _)

Ⓐ 12, 16, 25
Ⓑ 2, 6, 15
Ⓒ 35, 55, 100
Ⓓ 3, 5, 14

21. Which rule matches the pattern below?
0, 3, 6, 9, 12

Ⓐ **Multiply by 3**
Ⓑ **Add 3**
Ⓒ **Count up**
Ⓓ **0 through 12**

22. Which best represents the y-values in the chart?

x	y
0	2
3	8
6	11
9	20
12	26

Ⓐ **y is equal to x plus two**
Ⓑ **y is equal to x squared minus one**
Ⓒ **y is equal to two more than x times two**
Ⓓ **y is equal to two times x minus one**

23. Which is true about the x- and y-values in the chart?

x	y
0	2
3	8
6	14
9	20
12	26

Ⓐ **The y-values increase while the x-values decrease.**
Ⓑ **The x- and y-values increase at the same rate.**
Ⓒ **The x-values increase at a greater rate than the y-values.**
Ⓓ **The y-values increase at a greater rate than the x-values.**

24. Which coordinate pair would be plotted on the same line as the coordinates in the table below?

x	y
0	2
3	8
6	14
9	20
12	26

Ⓐ (5 , 12)
Ⓑ (2 , 4)
Ⓒ (10 , 23)
Ⓓ (7 , 18)

25. Which graph best represents the values in this table?

x	y
0	2
3	8
6	14
9	20
12	26

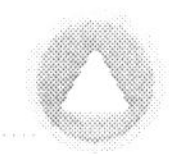

Ⓐ 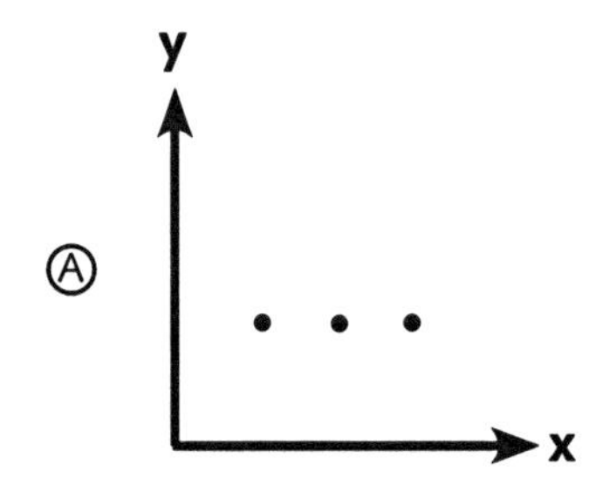

Ⓑ 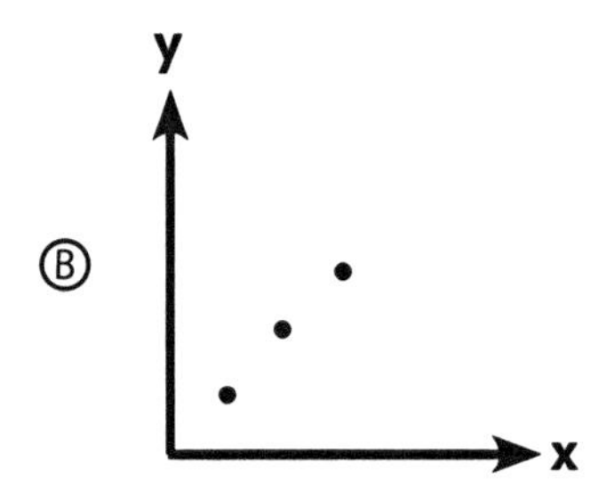

Ⓒ 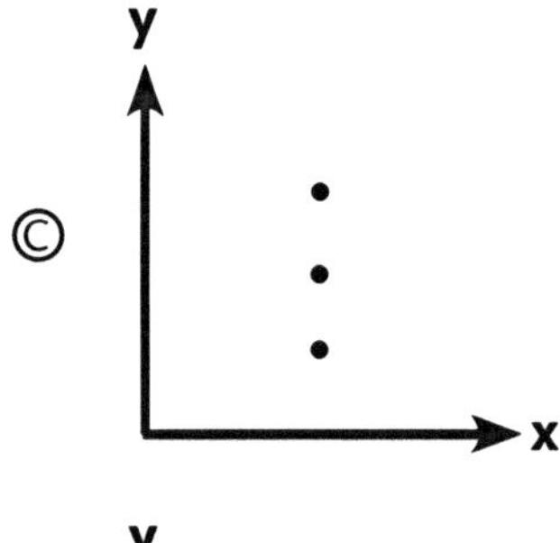

Ⓓ 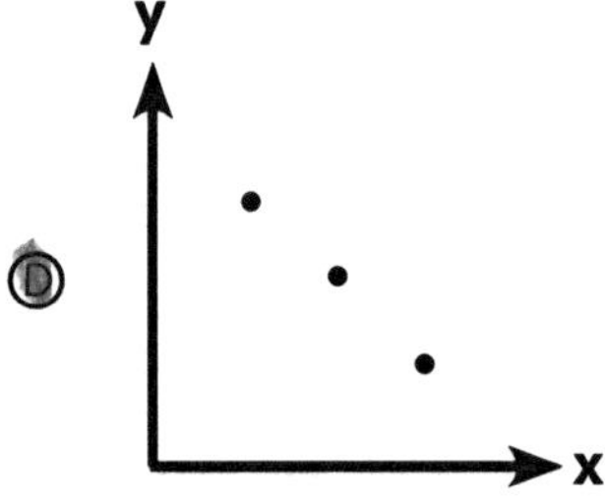

26. Which rule matches the pattern below?
16, 8, 4, 2, 1

Ⓐ **Subtract ½**
Ⓑ **Skip count backward**
Ⓒ **Even numbers**
Ⓓ **Divide by 2**

27. Which best represents the y-values in the chart?

x	y
16	255
8	63
4	15
2	3
1	0

Ⓐ **y is equal to x^2 minus one**
Ⓑ **y is equal to x times x**
Ⓒ **y is equal to one less than two times x**
Ⓓ **y is not related to x**

28. Which is true about the x- and y-values in the chart?

x	y
16	255
8	63
4	15
2	3
1	0

Ⓐ **The y-values increase while the x-values decrease.**
Ⓑ **The x- and y-values decrease at the same rate.**
Ⓒ **The y-values decrease at a greater rate than the x-values.**
Ⓓ **The x-values decrease at a greater rate than the y-values.**

29. Which coordinate pair would be plotted on the same line as the coordinates in the table below?

x	y
16	255
8	63
4	15
2	3
1	0

Ⓐ (17 , 256)
Ⓑ (7 , 48)
Ⓒ (3 , 9)
Ⓓ (12 , 145)

30. Which graph best represents a function in which y does not change?

Ⓐ
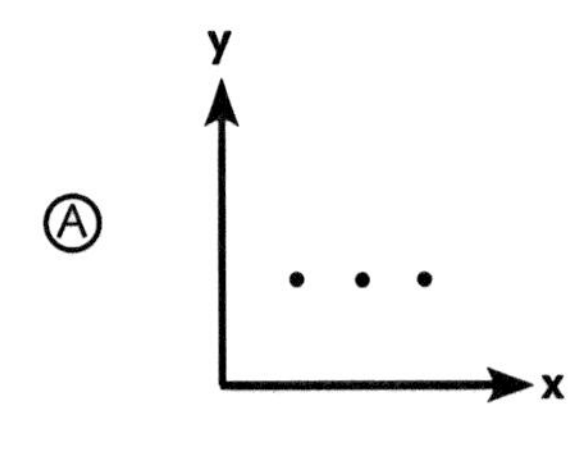

Ⓑ
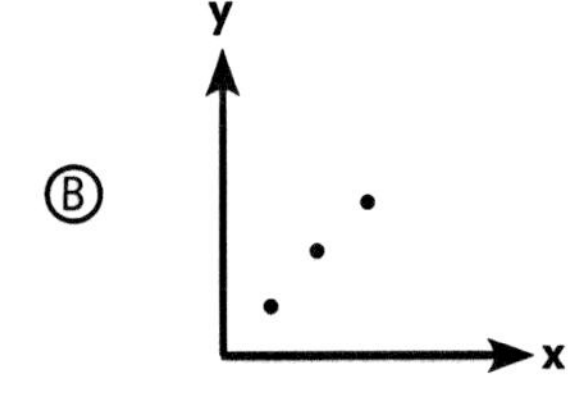

Ⓒ
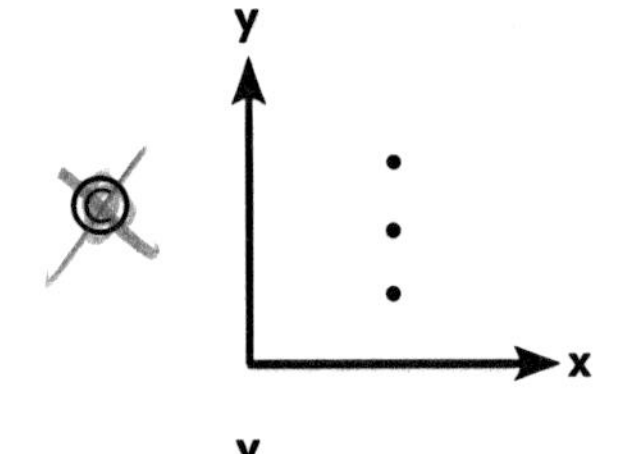

Ⓓ
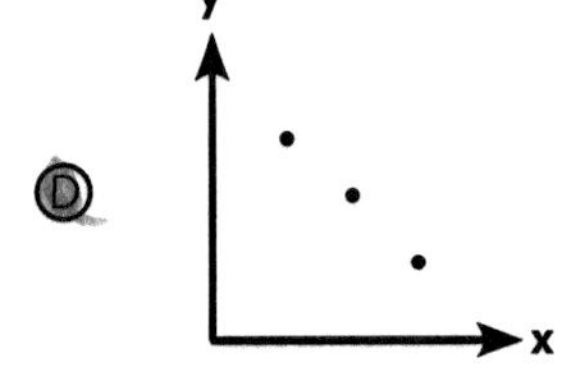

End of Operations and Algebraic Thinking

Operations and Algebraic Thinking

Answer Key
&
Detailed Explanations

Write and Interpret Numerical Expressions (5.OA.A.1)

Question No.	Answer	Detailed Explanation
1	D	Five and x are two addends that must equal 13. The only number that can be added to 5 to equal 13 is the number 8.
2	C	Decimals are read as one complete number (forty-five) followed by the place value of the digit furthest to the right (thousandths).
3	D	All of these situations would be represented with a negative number, because they are less than zero. A thermometer has 0 marked right on it. In football, the line of scrimmage is like the 0 mark and moving backwards would be negative yardage. Sea level is considered the 0 mark for altitude; anything deeper would be a negative value.
4	A	The words "shared equally" indicate division. In order to share the CDs equally among the people, divide the number of CDs (72) by the number of people (9) as shown by the equation $72 \div 9 = n$.
5	B	The words "gave her some more" indicate a missing addend. Julie started with some teddy bears (25), then she was given some more (+ m), now she has a new total (32). The correct equation is $25 + m = 32$.
6	A	In this equation, 7 plus some more equals a total of 21. Plugging the number 14 into the equation shows that $7 + 14 = 21$.
7	C	The numbers in the table do not follow a simple addition or multiplication rule (such as $x + 5 = y$ or $2x = y$), so it must be a combination of operations. Some guess-and-check shows that the numbers follow the pattern $3x - 2 = y$. For example, $3 (4) - 2 = 10$. Plugging in 10 for x, $3 (10) - 2 = y$. Therefore, $y = 28$.
8	D	The millions (79) are separated from the thousands (417) by a comma. Another comma separates the thousands from the hundreds (608). The first option has the correct place values but contains a number reversal on seventeen thousand.

Question No.	Answer	Detailed Explanation
9	B	The Associative Property of Multiplication states that when three or more numbers are multiplied, the product will be the same no matter how the three numbers are grouped. In this example, multiplying 11 x 6 x 7 will produce the same result whether the 11 x 6 are grouped together in parentheses or the 6 x 7 are grouped together. The other options are all mathematically correct, but they show different properties of multiplication.
10	A	By plugging in the values for x, the three options become: 405 - 100 > 300 or 305 > 300 405 - 105 > 300 or 300 > 300 405 - 110 > 300 or 295 > 300 The only option that produces a number greater than 300 is the first option, 100.
11	D	The equation mn ≥ 36 means that the product of m and n is greater than or equal to 36. Multiplying the two numbers given for m and n will result in the products 40, 36, and 36. All three of these products are greater than or equal to 36.
12	B	In this equation, p must be greater than 85 in order to subtract that amount and have 90 left over. This rules out the options 5 and 15. Plugging the first two options into the equation produces: 185 - 85 = 90 (incorrect) 175 - 85 = 90 (correct)
13	B	Fifteen is being divided by one of its factors to result in 5 (the other factor). The factor pair of 15 that includes a 5 is the pair 5 x 3. Check the equation by plugging in 3 for a: 15 ÷ 3 = 5.
14	A	In this equation, n is the number of people who shopped on Thursday. This value will be less than 120, because 120 is twice (more than) the number that shopped on Thursday. n would have to double to equal 120, so 2 x n = 120. The correct answer option shows this equation as 120 = n x 2.
15	C	The words "gave some to his sister" indicate a subtraction problem. Paul started with some crayons (52), then he gave some away (- c), now he is left with fewer (38). As an equation, this is written 52 - c = 38.

Question No.	Answer	Detailed Explanation
16	C	Before finding 9 less than the triple of x, determine the triple of x. In this example, 3x is 3 (16) = 48. Now, in order to find 9 less than the triple of x, subtract 9 from 48. 48 - 9 = 39.
17	B	The rule "y is equal to one less than the square of x" can be represented by the equation $y = x^2 - 1$. Plugging in 80 for y gives us $80 = x^2 - 1$. Therefore $81 = x^2$. Remember that x^2 means x times itself. The only number that works here is 9 (9 x 9 = 81).
18	C	The first step in the pattern, 32 + x = 44, gives us a rule Add 12. To check this answer, apply the rule to each number in the pattern. 44 + 12 = 56 56 + 12 = 68
19	A	The population increases by 1,000 people per year. In five years, the population should increase by 5,000. There were 40,000 people in 2010, so there should be 5,000 more, or 45,000 people, in 2015.
20	D	The first week, Kevin earned $10.00. The second week, he earned twice that, or $20.00. The third week, he earned twice that, or $40.00. The fourth week, he earned twice that, or $80.00. To find the total for all four weeks, add $10.00 + $20.00 + $40.00 + $80.00 to get a total of $150.00.

Record and Interpret Calculations with Numbers (5.OA.A.2)

Question No.	Answer	Detailed Explanation
1	B	First, find the quotient of 72 divided by 8 (72 ÷ 8). Then determine what ten more than that would be (+ 10).
2	D	First, find the product of 12 and 4 (12 x 4). Then subtract this amount from 75.
3	A	Show 10 cases of 24 as (10 x 24) and 3 six-packs as (3 x 6). Add the two expressions to find the total: (10 x 24) + (3 x 6).
4	B	First, subtract the 9 that Olivia kept for herself (42 – 9). Then divide the difference among the three friends: (42 – 9) ÷ 3.
5	C	The expression 8 x (467 + 509) indicates that you should first find the sum of 467 and 509, and then multiply by 8. Therefore, the solution is 8 times greater than that sum.

Question No.	Answer	Detailed Explanation
6	A	The expression (3,258 – 741) ÷ 3 indicates that you should first find the difference of 3,258 and 741, and then divide by 3. Therefore, the solution is one third as much as that difference.
7	C	A quick estimate shows that option C, a number in the hundreds times a number in the tens, would result in a number in the thousands. The other options would all result in a number in the hundreds.
8	D	A quick estimate shows that option D, in which the largest amount is subtracted from 684, would result in the smallest number. Options A and B subtract a relatively small amount from 684, and option C will actually result in a larger number.
9	B	First, multiply 7 items by the 25 students (25 x 7). Then add to that product the 13 the teacher sold: 13 + (25 x 7).
10	A	First, double the $75 he had (75 x 2). Then subtract the $47 he spent: (75 x 2) - 47.

Analyze Patterns and Relationships (5.OA.B.3)

Question No.	Answer	Detailed Explanation
1	B	The rule is x3, so plugging in each input number results in the following: 1 x 3 = 3, 2 x 3 = 6, 8 x 3 = 24, 12 x 3 = 36.
2	C	The rule is +4, ÷2, so plugging in each input number results in the following: (4 + 4) ÷ 2 = 4, (6 + 4) ÷ 2 = 5, (22 + 4) ÷ 2 = 13, (40 + 4) ÷ 2 = 22.
3	A	The rule is x2, -1, so plugging in each input number results in the following: 5 x 2 - 1 = 9, 9 x 2 - 1 = 17, 14 x 2 - 1 = 27, 25 x 2 - 49 = 9. To create coordinate pairs, write the input number followed by the output number, separated by a comma, in parentheses.
4	B	The rule is ÷3, +2, so plugging in each input number results in the following: 9 ÷ 3 + 2 = 5, 15 ÷ 3 + 2 = 7, 27 ÷ 3 + 2 = 11, 33 ÷ 3 + 2 = 13. To create coordinate pairs, write the input number followed by the output number, separated by a comma, in parentheses.

Question No.	Answer	Detailed Explanation
5	D	The rule is -4, so plugging in each output number results in the following: 5 – 4 = 1, 11 – 4 = 7, 14 – 4 = 10, 19 – 4 = 15.
6	D	The rule is +1, x5, so plugging in each output number results in the following: (0 + 1) x 5 = 5, (3 + 1) x 5 = 20, (6 + 1) x 5 = 35, (10 + 1) x 5 = 55.
7	B	Subtracting 6 from each x-value results in the corresponding y-value as follows: 11 – 6 = 5, 14 – 6 = 8, 21 – 6 = 15, 28 – 6 = 22.
8	A	Multiplying each x-value by 2 and then subtracting 4 results in the corresponding y-value as follows: 4 x 2 – 4 = 4, 7 x 2 – 4 = 10, 13 x 2 – 4 = 22, 20 x 2 – 4 = 36.
9	C	As the value of x increases, the value of y increases, both at fixed rates. This produces an upward-sloping straight line.
10	D	If the value of x increases while the value of y decreases, the function produces a downward sloping straight line.
11	C	According to the rules, sequence x would begin 2, 5, 8, ..., and sequence y would begin 4, 10, 16, ... Therefore, the terms in sequence y are always two times the terms in sequence x.
12	B	According to the rules, sequence x would begin 1, 2, 4, ..., and sequence y would begin 2, 4, 8, ... Therefore, the terms in sequence y are always two times the terms in sequence x.
13	A	According to the rule in option A, sequence x would begin 5, 11, 17, ..., and sequence y would begin 8, 14, 20, ... Therefore, the terms in sequence y are always three more than the terms in sequence x.
14	D	According to the rule in option D, sequence x would begin 4, 8, 16, ..., and sequence y would begin 2, 4, 8, ... Therefore, the terms in sequence y are always half as much as the terms in sequence x.

Question No.	Answer	Detailed Explanation
15	A	The rule "subtract 5" can be represented as x – 5 = y. Therefore, plugging in the x- and y-values from the coordinate pairs results in the following true expressions: 12 – 5 = 7, 14 – 5 = 9, 18 – 5 = 13.
16	C	The rule x2, +3 can be represented as 2x + 3 = y. Therefore, plugging in the x- and y-values from the coordinate pairs results in the following true expressions: 2(3) + 3 = 9, 2(5) + 3 = 13, 2(8) + 3 = 19.
17	C	The relationship "x is always 3 less than y" can be represented as x + 3 = y. Therefore, plugging in the x- and y-values from the coordinate pairs results in the following true expressions: 4 + 3 = 7, 9 + 3 = 12, 11 + 3 = 14.
18	D	The relationship "y is always 4 times as much as x" can be represented as 4x = y. Therefore, plugging in the x- and y-values from the coordinate pairs results in the following true expressions: 4(2) = 8, 4(5) = 20, 4(10) = 40.
19	B	Plugging the x-values into the rule y = 3x + 2 results in the following values for y: 5 = 3(1) + 2, 14 = 3(4) + 2, 26 = 3(8) + 2.
20	B	Plugging the x-values into the rule x = y + 5 results in the following values for y: 7 = 2 + 5, 11 = 6 + 5, 20 = 15 + 5.
21	B	Beginning with 0, adding three would result in 3. Adding 3 more would result in 6, and so on.
22	C	The statement y is equal to two more than x times two can be checked with the formula y = 2x + 2. Plugging in the coordinates gives us: 2 = 2(0) + 2, 8 = 2(3) + 2, 14 = 2(6) + 2, 20 = 2(9) + 2, 26 = 2(10) + 2.
23	D	The y-values increase by 6 each term, while the x-values increase by 3. Therefore, the y-values increase at a greater rate than the x-values.
24	A	Since the chart follows the pattern y= 2x + 2, plugging in the coordinates (5 , 12) gives us 12 = 10 + 2, which is true.

Question No.	Answer	Detailed Explanation
25	B	As the value of x increases, the value of y increases, both at fixed rates. This produces an upward-sloping straight line.
26	D	Beginning with 16, dividing by 2 would result in 8. Dividing that by 2 would result in 4, and so on. Subtracting ½ from 16 would result in 15 ½.
27	A	The statement y is equal to x^2 minus one can be checked with the formula $y = x^2 - 1$. Plugging in the coordinates (4 , 15) gives us: 255 = 162 - 1, 63 = 82 - 1, 15 = 42 - 1, 3 = 22 - 1, 0 = 12 - 1
28	C	The x-values decrease in half each term, while the y-values decrease by more than half. Therefore, the y-values decrease at a greater rate than the x-values.
29	B	Since the chart follows the pattern $y= x^2 - 1$, plugging in the coordinates (7 , 48) gives us 48 = 49 - 1, which is true.
30	A	A horizontal line shows increasing values for x with no change in the value for y.

Name: ______________________ Date: ______________

Numbers and Operations in Base Ten

Place Value (5.NBT.A.1)

1. In the number 913,874 which digit is in the ten thousands place?

Ⓐ 8
Ⓑ 1
Ⓒ 9
Ⓓ 3

2. In the number 7.2065 which digit is in the thousandths place?

Ⓐ 5
Ⓑ 2
Ⓒ 0
Ⓓ 6

3. Which number is equivalent to 8/10?

Ⓐ 0.8
Ⓑ 8.0
Ⓒ 0.08
Ⓓ 0.008

4. What is the equivalent of 4 and 3/100?

Ⓐ 40.3
Ⓑ 0.403
Ⓒ 4.03
Ⓓ 403.0

5. In the number 16,428,095 what is the value of the digit 6?

Ⓐ 6 million
Ⓑ 60 thousand
Ⓒ 16 million
Ⓓ 600 thousand

6. What is the value of 9 in the number 5,802.109

Ⓐ 9 thousand
Ⓑ 9 tenths
Ⓒ 9 thousandths
Ⓓ 9 hundredths

7. Which comparison is correct?

Ⓐ 50.5 = 50.05
Ⓑ 0.05 = 0.50
Ⓒ 0.005 = 500.0
Ⓓ 0.50 = 0.500

8. Which number is one hundredth less than 406.51?

Ⓐ 406.41
Ⓑ 406.50
Ⓒ 306.51
Ⓓ 406.01

9. Which of the following numbers is greater than 8.4?

Ⓐ 8.41
Ⓑ 8.40
Ⓒ 8.14
Ⓓ 8.04

10. Which of the following numbers is less than 2.17?

Ⓐ 21.7
Ⓑ 2.71
Ⓒ 2.170
Ⓓ 2.07

Name: ______________________ Date: ______________

Multiplication & Division of Powers of Ten (5.NBT.A.2)

1. **Solve: 9×10^3 =**

   Ⓐ 900
   Ⓑ 9,000
   Ⓒ 117
   Ⓓ 270

2. **What is the quotient of 10^7 divided by 100?**

   Ⓐ 0.7
   Ⓑ 100,000
   Ⓒ 70,000
   Ⓓ 700

3. **Solve: 0.51 x ____ = 5,100**

   Ⓐ 10^4
   Ⓑ 10^2
   Ⓒ 100
   Ⓓ 10^3

4. **Astronomers calculate a distant star to be 3×10^5 light years away. How far away is the star?**

   Ⓐ 30,000 light years
   Ⓑ 3,000 light years
   Ⓒ 3,000,000 light years
   Ⓓ 300,000 light years

5. **A scientist calculates the weight of a substance as $6.9 \div 10^4$ grams. What is the weight of the substance?**

   Ⓐ 69,000 grams
   Ⓑ 69 milligrams
   Ⓒ 0.00069 grams
   Ⓓ 6.9 kilograms

6. **Looking through a microscope, a doctor finds a germ that is 0.00000082 millimeters long. How can he write this number in his notes?**

   Ⓐ 8.2×10^7
   Ⓑ $8.2 \div 10^7$
   Ⓒ $8.2 \times 10^{0.00001}$
   Ⓓ $8.2 \div 700$

7. **Which of the following is 10^5 times greater than 0.016?**

Ⓐ 160
Ⓑ 1,600
Ⓒ 16.0
Ⓓ 1.60

8. **Find the missing number.**
_____ x 477 = 47,700,000

Ⓐ 10,000,000
Ⓑ 10 x 5
Ⓒ 10^5
Ⓓ 1,000

9. **The number 113 is _____ times greater than 0.113.**

Ⓐ 3
Ⓑ 10^3
Ⓒ 30
Ⓓ 10,000

10. **________ ÷ 10^5 = 4.6**

Ⓐ 460
Ⓑ 4,600
Ⓒ 46,000
Ⓓ 460,000

Read and Write Decimals (5.NBT.A.3.A)

1. How is the number four hundredths written?

   Ⓐ 0.04
   Ⓑ 0.400
   Ⓒ 400.0
   Ⓓ 0.004

2. How is the number 0.2 read?

   Ⓐ Zero and two
   Ⓑ Decimal two
   Ⓒ Two tenths
   Ⓓ Two hundredths

3. What is the decimal form of 7/10?

   Ⓐ 7.10
   Ⓑ 0.7
   Ⓒ 10.7
   Ⓓ 0.07

4. The number 0.05 can be represented by which fraction?

   Ⓐ 0/5
   Ⓑ 5/100
   Ⓒ 5/10
   Ⓓ 1/05

5. Which of the following numbers is equivalent to one half?

   Ⓐ 0.2
   Ⓑ 0.12
   Ⓒ 1.2
   Ⓓ 0.5

6. How is the number sixty three hundredths written?

   Ⓐ 0.63
   Ⓑ 0.063
   Ⓒ 0.0063
   Ⓓ 6.300

7. **What is the correct way to read the number 40.057?**

Ⓐ **Forty point five seven**
Ⓑ **Forty and fifty-seven hundredths**
Ⓒ **Forty and fifty-seven thousandths**
Ⓓ **Forty and five hundredths and seven thousandths**

8. **Which of the following numbers has:**
0 in the hundredths place
8 in the tenths place
3 in the thousandths place
9 in the ones place

Ⓐ **9.083**
Ⓑ **0.839**
Ⓒ **9.803**
Ⓓ **0.9803**

9. **For which number is this the expanded form?**
9 x 10 + 2 x 1 + 3 x (1/10) + 8 x (1/100)

Ⓐ **98.08**
Ⓑ **93.48**
Ⓒ **9.238**
Ⓓ **92.38**

10. **What is the correct expanded form of the number 0.85?**

Ⓐ **(8 x 10) + (5 x 100)**
Ⓑ **8 x (1/10) + 5 x (1/100)**
Ⓒ **85 ÷ 10**
Ⓓ **(8 ÷ 10) x (5 ÷ 10)**

Name: ________________________ Date: ____________

Comparing & Ordering Decimals (5.NBT.A.3.B)

1. **Which of the following numbers is the least?**
0.04, 4.00, 0.40, 40.0

Ⓐ 0.04
Ⓑ 4.00
Ⓒ 0.40
Ⓓ 40.0

2. **Which of the following numbers is greatest?**
0.125, 0.251, 0.512, 0.215

Ⓐ 0.125
Ⓑ 0.251
Ⓒ 0.512
Ⓓ 0.215

3. **Which of the following numbers is less than seven hundredths?**

Ⓐ 0.072
Ⓑ 0.60
Ⓒ 0.058
Ⓓ All of these

4. **Which of the following comparisons is correct?**

Ⓐ 48.01 = 48.1
Ⓑ 25.4 < 25.40
Ⓒ 10.83 < 10.093
Ⓓ 392.01 < 392.1

5. **Arrange these numbers in order from least to greatest:**
1.02, 1.2, 1.12, 2.12

Ⓐ 1.2, 1.12, 1.02, 2.12
Ⓑ 2.12, 1.2, 1.12, 1.02
Ⓒ 1.02, 1.12, 1.2, 2.12
Ⓓ 1.12, 2.12, 1.02, 1.2

6. **Which of the following is true?**

Ⓐ 3.21 > 32.1
Ⓑ 32.12 > 312.12
Ⓒ 32.12 > 3.212
Ⓓ 212.3 < 21.32

7. **Arrange these numbers in order from greatest to least:**
   2.4, 2.04, 2.21, 2.20

   Ⓐ 2.4, 2.04, 2.21, 2.20
   Ⓑ 2.4, 2.21, 2.20, 2.04
   Ⓒ 2.21, 2.20, 2.4, 2.04
   Ⓓ 2.20, 2.4, 2.04, 2.21

8. **Which of the following numbers completes the sequence below?**
   4.17, _____, 4.19

   Ⓐ 4.18
   Ⓑ 4.81
   Ⓒ 5.17
   Ⓓ 4.27

9. **Which of the following comparisons is true?**

   Ⓐ 0.403 > 0.304
   Ⓑ 0.043 < 0.403
   Ⓒ 0.043 < 0.304
   Ⓓ All of the above

10. **Which number completes the following sequence?**
    2.038, 2.039, _______

    Ⓐ 2.049
    Ⓑ 2.400
    Ⓒ 2.0391
    Ⓓ 2.04

Name: ______________________ Date: ______________

Rounding Decimals (5.NBT.A.4)

1. Is $7.48 closest to $6, $7 or $8?

Ⓐ $6
Ⓑ $7
Ⓒ $8
Ⓓ It is right in the middle of $7 and $8

2. Round the Olympic time of 56.389 to the nearest tenth of a second.

Ⓐ 56.0
Ⓑ 57
Ⓒ 56.4
Ⓓ 56.39

3. Round the number 57.81492 to the nearest hundredth.

Ⓐ 57.82
Ⓑ 58.00
Ⓒ 57.80
Ⓓ 57.81

4. Which of the following numbers would round to 13.75?

Ⓐ 13.755
Ⓑ 13.70
Ⓒ 13.756
Ⓓ 13.747

5. Jerry spent $5.91, $7.27, and $12.60 on breakfast, lunch, and dinner. About how much did his meals cost in all?

Ⓐ about $24
Ⓑ about $26
Ⓒ about $25
Ⓓ about $27

6. Maria needs to buy wood for a door frame. She needs two pieces that are 6.21 feet long and one piece that is 2.5 feet long. About how much wood should she buy?

Ⓐ about 15 feet
Ⓑ about 9 feet
Ⓒ about 17 feet
Ⓓ about 14 feet

7. **Mika has a rectangular flower garden. It measures 12.2 meters on one side and 7.8 meters on the other. What is a reasonable estimation of the area of the flower garden? (Area= length x width)**

Ⓐ 96 square meters
Ⓑ 20 square meters
Ⓒ 66 square meters
Ⓓ 120 square meters

8. **Shanda ran a lap in 6.78 minutes. Assuming she maintains this time for every lap she runs, estimate the time it would take her to run three laps.**

Ⓐ 25 minutes
Ⓑ 21 minutes
Ⓒ 18 minutes
Ⓓ 10 minutes

9. **A basketball player scores an average of 13.2 points per game. During a 62-game season, he would be expected to score about ________ points. (Assume he will play every game.)**

Ⓐ 600 points
Ⓑ 1,000 points
Ⓒ 800 points
Ⓓ 400 points

10. **Use estimation to complete the following:**
The difference of 31.245 - 1.396 is between _____.

Ⓐ 29 and 29.5
Ⓑ 29.5 and 30
Ⓒ 30 and 30.5
Ⓓ 30.5 and 31

Name: ______________________ Date: ______________

Multiplication of Whole Numbers (5.NBT.B.5)

1. **Solve. 79 x 14 = ____**

   Ⓐ 790
   Ⓑ 1,106
   Ⓒ 854
   Ⓓ 224

2. **A farmer plants 18 rows of beans. If there are 50 bean plants in each row, how many plants will he have altogether?**

   Ⓐ 908
   Ⓑ 68
   Ⓒ 900
   Ⓓ 98

3. **Solve. 680 x 94 = _____**

   Ⓐ 64,070
   Ⓑ 63,960
   Ⓒ 64,760
   Ⓓ 63,920

4. **What is the missing value?**
   ____ x 11 = 374

   Ⓐ 36
   Ⓑ 30
   Ⓒ 34
   Ⓓ 31

5. **Which of the following statements is true?**

   Ⓐ 28 x 17 = 17 x 28
   Ⓑ 28 x 17 = 20 x 8 x 10 x 7
   Ⓒ 28 x 17 = (28 x 1) + (28 x 7)
   Ⓓ 28 x 17 = 27 x 18

6. **Which equation is represented by this array?**

Ⓐ 3 + 7 + 3 + 7 = 20
Ⓑ 7 + 7 + 7 + 7 + 7 = 35
Ⓒ 3 x 3 + 7 = 16
Ⓓ 3 x 7 = 21

7. **What would be a quick way to solve 596 x 101 accurately?**

Ⓐ Multiply 5 x 101, 9 x 101, 6 x 101, then add the products.
Ⓑ Multiply 596 x 100 then add 596 more.
Ⓒ Shift the 1 and multiply 597 x 100 instead.
Ⓓ Estimate 600 x 100.

8. **Harold baked 9 trays of cookies for a party. Three of the trays held 15 cookies each and six of the trays held 18 cookies each. How many cookies did Harold bake in all?**

Ⓐ 297
Ⓑ 135
Ⓒ 153
Ⓓ 162

9. **What's wrong with the following computation?**

```
    28
   x53
  ----
    32
    60
   400
 +1000
 -----
  1492
```

Ⓐ 3 x 8 is multiplied incorrectly.
Ⓑ 50 x 20 should only have two zeros.
Ⓒ 5 x 8 is only 40.
Ⓓ There's a missing 1 that should have been carried from the tens to the hundreds place.

10. Solve.
407 x 35 = _____

Ⓐ **14,280**
Ⓑ **14,245**
Ⓒ **12,445**
Ⓓ **16,135**

Division of Whole Numbers (5.NBT.B.6)

1. **Find the missing number:**
   48 ÷ ___ = 12

   Ⓐ **4**
   Ⓑ **10**
   Ⓒ **6**
   Ⓓ **8**

2. **Hannah is filling gift bags for a party. She has 72 pieces of candy to pass out. If there are 8 bags, how many pieces of candy will go in each bag?**

   Ⓐ **8 R 5**
   Ⓑ **10**
   Ⓒ **9**
   Ⓓ **7**

3. **Solve. 1,248÷ 6 =**

   Ⓐ **2,080**
   Ⓑ **208**
   Ⓒ **28**
   Ⓓ **280**

4. **The fifth grade class took a field trip to the theater. The 96 students sat in rows with 10 students in each row. How many rows did they use?**

   Ⓐ **11**
   Ⓑ **9**
   Ⓒ **10**
   Ⓓ **12**

5. **What is the value of 6,720 ÷ 15?**

   Ⓐ **510**
   Ⓑ **426**
   Ⓒ **448**
   Ⓓ **528**

6 **What is 675,000 divided by 100?**

   Ⓐ **675**
   Ⓑ **67,500**
   Ⓒ **67.5**
   Ⓓ **6,750**

7. Which of the following statements is true?

Ⓐ 75 ÷ 0 = 0
Ⓑ 75 ÷ 0 = 1
Ⓒ 75 ÷ 0 = 75
Ⓓ 75 ÷ 0 cannot be solved

8. Taylor is putting 100 donuts into boxes. Each box holds 12 donuts. How many donuts will be left over after filling each box?

Ⓐ 4
Ⓑ 8
Ⓒ 9
Ⓓ 5

9. Which of the following statements is true?

Ⓐ 26 ÷ 1 = 1
Ⓑ 26 ÷ 1 = 26
Ⓒ 26 ÷ 1 = 0
Ⓓ 26 ÷ 1 cannot be solved

10. Jeremy is rolling coins to take to the bank. He has 680 nickels to roll. If each sleeve holds 40 nickels, how many sleeves will he be able to fill?

Ⓐ 8
Ⓑ 17
Ⓒ 16
Ⓓ 12

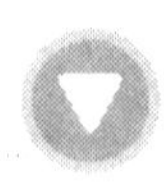

Add, Subtract, Multiply, and Divide Decimals (5.NBT.B.7)

1. At a math competition, three members of a team each solved a problem as quickly as they could. Their times were 4.18 seconds, 3.75 seconds, and 3.99 seconds. What was the total of their three times?

Ⓐ 11.92 seconds
Ⓑ 10.99 seconds
Ⓒ 10.72 seconds
Ⓓ 11.72 seconds

2. Beginning with the number 6.472, add:
1 hundredth
3 ones
5 tenths
What is the result?

Ⓐ 7.822
Ⓑ 6.823
Ⓒ 9.982
Ⓓ 6.607

3. Find the perimeter (total length of all four sides) of a trapezoid whose sides measure 2.09 ft, 2.09 ft, 3.72 ft, and 6.60 ft.

Ⓐ 16.12 ft
Ⓑ 14.5 ft
Ⓒ 13.50 ft
Ⓓ 8.56 ft

4. Find the difference:
85.37 - 75.2 =

Ⓐ 160.57
Ⓑ 10
Ⓒ 10.35
Ⓓ 10.17

5. Subtract:
3.64 - 1.46 =

Ⓐ 2.18
Ⓑ 4.18
Ⓒ 1.18
Ⓓ 4.20

6. **Normal body temperature is 98.6 degrees Fahrenheit. When Tyler had a fever, his temperature went up to 102.2 degrees. By how much did Tyler's temperature increase?**

Ⓐ 4.4 degrees
Ⓑ 3.6 degrees
Ⓒ 4.2 degrees
Ⓓ 3.2 degrees

7. **A stamp costs $0.42. How much money would you need to buy 8 stamps?**

Ⓐ $.82
Ⓑ $3.33
Ⓒ $3.36
Ⓓ $4.52

8. **Find the product:**
0.25 x 1.1 =

Ⓐ .75
Ⓑ 0.275
Ⓒ 0.27
Ⓓ .25

9. **Divide 0.42 by 3.**

Ⓐ 14
Ⓑ 126
Ⓒ 0.14
Ⓓ 12.6

10. **Solve:**
0.09 ÷ 0.3 =

Ⓐ 0.27
Ⓑ 0.003
Ⓒ 0.027
Ⓓ 0.3

End of Numbers and Operations in Base Ten

Numbers and Operations in Base Ten

Answer Key & Detailed Explanations

Place Value (5.NBT.A.1)

Question No.	Answer	Detailed Explanation
1	B	The ten thousands place is five places to the left of the decimal, so the 1 is in the ten thousands place.
2	D	The thousandths place is three places to the right of the decimal, so the 6 is in the thousandths place.
3	A	The tenths place is immediately to the right of the decimal. In order to show eight-tenths, use an 8 immediately to the right of the decimal. It is common to use a place-holder 0 in the ones place.
4	C	Write the number 4 in the ones place. The word 'and' indicates the decimal point. The fractional part of the number is three-hundredths, which is shown with a 3 in the hundredths place. Use a place holder 0 in the tenths place, so the 3 is two places to the right of the decimal.
5	A	The 6 is seven places to the left of the decimal, which is the millions place. Its value is 6 million.
6	C	The 9 is three places to the right of the decimal, which is the thousandths place. Its value is 9 thousandths.
7	D	In order for two numbers to be equal, they must have the same digits in the same place value. In this option, each number has a 5 in the tenths place. The final zeros after the tenths place do not change the value.
8	B	The hundredths place is two places to the right of the decimal. There is a 1 in the hundredths place, so one hundredth less would be 0, making the number 406.50.
9	A	The number 8.4 can be thought of as 8.40 (the final zero does not change the value). In this case, the number 8.41 would be greater because there is 1 hundredth compared to 0 hundredths. The other options are incorrect because they are equal to or less than 8.4 because the digit in the tenths place is lower.

Question No.	Answer	Detailed Explanation
10	D	Looking to the right of the decimal, every option is greater than or equal to 2.17. The only option with a smaller digit in the tenths place is 2.07.

Multiplication & Division of Powers of Ten (5.NBT.A.2)

Question No.	Answer	Detailed Explanation
1	B	10^3 means 10 x 10 x 10, which equals 1,000. 9 x 1,000 = 9,000. Another way to think of this problem is 9 x 10 = 90, then make sure the number of zeros in the answer matches the number of the exponent (3), which is 9,000.
2	B	10^7 means 10 x 10 x 10 x 10 x 10 x 10 x 10, which equals 10,000,000. Divide 10,000,000 by 100, or move the decimal point to the left (because it is division) two places to get 100,000.
3	A	Since the decimal point in 0.51 is being moved four places to the right, it is being multiplied by 10,000. This number can be shown as 10^4.
4	D	10^5 means 10 x 10 x 10 x 10 x 10, which equals 100,000. 3 x 100,000 = 300,000. Another way to think of this problem is 3 x 10 = 30, then make sure the number of zeros in the answer matches the number of the exponent (5), which is 300,000.
5	C	10^4 means 10 x 10 x 10 x 10, which equals 10,000. 6.9 ÷ 10,000 = 0.00069. Another way to think of this problem is to move the decimal in 6.9 to the left (because it is division) the number of places equal to the exponent (4).
6	B	The decimal point is being moved to the left, so it is a division problem. Since it is being moved 7 places, 8.2 is being divided by 10^7.
7	B	10^5 means 10 x 10 x 10 x 10 x 10, which equals 100,000. To find the number that is 100,000 times greater than 0.016, multiply 0.016 x 100,000, or move the decimal point to the right (because it is multiplication) the same number of places as the exponent (5).
8	C	Since the decimal point in 477.0 is being moved five places to the right, it is being multiplied by 100,000. This number can be shown as 10^5.

Question No.	Answer	Detailed Explanation
9	B	Since the decimal point in 113.0 is being moved three places to the left, it is being divided by 1,000. This number can be shown as 10^3.
10	D	10^5 means 10 x 10 x 10 x 10 x 10, which equals 100,000. Dividing by 100,000 is the same as moving the decimal place in five places to the left. Moving the decimal point in 4.6 five places back to the right would give you 460,000.

Read and Write Decimals (5.NBT.A.3.A)

Question No.	Answer	Detailed Explanation
1	A	The 4 goes in the hundredths place, which is two places to the right of the decimal. All other places get place-holder zeros.
2	C	The two is immediately to the right of the decimal, so it is in the tenths place. It is read "two tenths."
3	B	The fraction is seven tenths. To show this value in decimal form, use the digit 7 in the tenths place (immediately to the right of the decimal).
4	B	In the number 0.05, the 5 is in the hundredths place. To show this amount (five hundredths) as a fraction, use 5 as the numerator and 100 as the denominator.
5	D	One half is equal to five tenths (think of a pizza sliced into 10 pieces, half of the pizza would be 5 out of 10 slices). To show five tenths, use a 5 in the tenths place immediately to the right of the decimal.
6	A	Sixty hundredths is equivalent to six tenths (the place to the right of the decimal). Three hundredths is shown by a 3 in the hundredths place (two places to the right of the decimal).
7	C	Begin by saying the whole number (forty), the word 'and' for the decimal, and then the decimal portion of the number. The decimal .057 is fifty-seven thousandths. The 5 hundredths is equivalent to fifty thousandths.
8	C	The hundredths place (0) is two places to the right of the decimal. The tenths place (8) is immediately to the right of the decimal. The thousandths place (3) is three places to the right of the decimal. The ones place (9) is immediately to the left of the decimal.

Question No.	Answer	Detailed Explanation
9	D	In expanded form, each digit is multiplied by its place value and the products are added together. The expression 9 x 10 + 2 x 1 + 3 x (1/10) + 8 x (1/100) can be thought of as: 9 x 10 = 90 2 x 1 = 2 3 x (1/10) = .3 8 x (1/100) = .08 Add the products to get 92.38
10	B	In expanded form, each digit is multiplied by its place value and the products are added together. The number 0.85 is made up of 8 tenths, which is written 8 x (1/10), and 5 hundredths, which is written 5 x (1/100).

Comparing & Ordering Decimals (5.NBT.A.3.B)

Question No.	Answer	Detailed Explanation
1	A	Since each of the options contains only one non-zero digit (4), compare the place value of the 4 to find the lowest number. 0.04 is the lowest number because the 4 is the furthest to the right of the decimal (in the hundredths place).
2	C	In order to find the greatest number, compare the digit in the highest place value. All of the options have 0 ones, so look to the tenths place. The number with 5 in the tenths place is greater than the numbers with 1 or 2 in the tenths place, no matter what comes next.
3	C	Seven hundredths is written 0.07. In order for a number to be lower, it has to have 0 in the tenths place and a digit lower than 7 in the hundredths place.

Name: ______________________ Date: ______________

Question No.	Answer	Detailed Explanation
4	D	Each of these options involves comparing decimals (since the numbers to the left of the decimal point in each option are equal). Remember, the further a number is to the right of a decimal, the lower its place value. Be sure to compare numbers that are in the same place value (compare tenths to tenths, etc.). For each of these options, compare the underlined digit: 48.01 = 48.1 These are not equal, because .0 is less than .1. 25.4_< 25.40 The final zero does not affect the number's value, so these two numbers are equal. 10.83 < 10.093 The 8 in the tenths place is greater than a 0 in the tenths place. 392.01 < 392.1 This is correct because .1 is greater than .0.
5	C	In order to compare the size of numbers, begin with the place value furthest to the left. In this case, three of the numbers have a 1 in the ones place, so look to the tenths place to compare those three. The number with the lowest digit in the tenths place will come first (1.02) followed by the number with the next-highest digit in the tenths place (1.12) followed by the number with the highest digit in the tenths place (1.2). The remaining number has a 2 in the ones place, so it is the greatest.
6	C	Each of these options involves comparing similar digits in different place values. Be sure to compare numbers that are in the same place value (compare tenths to tenths, etc.), starting with the highest place value. For each of these options, compare the underlined digit: _3.21 > 32.1 No tens is less than 3 tens. _32.12 > 312.12 No hundreds is less than 3 hundreds. 32.12 > 3.212 This is correct because 3 tens is more than no tens. 212.3 < 21.32 Two hundreds is greater than no hundreds.

Question No.	Answer	Detailed Explanation
7	B	In order to compare the size of numbers, begin with the place value furthest to the left. In this case, all of the numbers have a 2 in the ones place, so look to the tenths place to compare them. The number with the highest digit in the tenths place will come first (2.4). The next two highest numbers both have a 2 in the tenths place, so look to the hundredths place. The number with the highest digit in the hundreds place will come first (2.21) followed by the number with the lower digit in the hundreds place (2.20). The final number has a 0 in the tenths place, so it is the least of all.
8	A	The missing number must be greater than 4.17 but less than 4.19. Since the ones place (4) and tenths place (1) are the same, the missing number will begin with 4.1 as well. Looking to the hundredths place, the missing number must fall between 7 and 9. That makes it 4.18.
9	D	To compare these numbers, look at the digit in the highest place value (the tenths place). 0.403 is greater than 0.304 and that's greater than 0.043.
10	D	This pattern is increasing by one thousandth every term. After 2.039, the thousandths place will increase by one. Since there is already a 9 in the thousandths place, it will become zero and the hundredths place will increase to 4. The number 2.040 can also be written 2.04.

Rounding Decimals (5.NBT.A.4)

Question No.	Answer	Detailed Explanation
1	B	In order to round to the nearest whole dollar, look to the tenths (dimes) place. In $7.48 there is a 4 in the tenths place, which means round down to $7.00.
2	C	In the number 56.389, there is a 3 in the tenths place. Look to the right to see that 8 means round up. The number becomes 56.4 with no hundredths or thousandths.
3	D	In the number 57.81492, there is a 1 in the hundredths place. To determine whether to round up to 2 or remain 1, look to the digit to the right. A 4 means that the number will round down to 57.81.

Question No.	Answer	Detailed Explanation
4	D	In order for a number to round to 13.75, it must be between 13.745 and 13.754. The number 13.747 has a 7 in the thousandths place that means it will round up to 13.75.
5	B	Round $5.91 up to $6. Round $7.27 down to $7. Round $12.60 up to $13. $6 + $7 + $13 = $26.
6	A	Round 6.21 down to 6 and 2.5 up to 3. She needs two 6-foot pieces (12 feet) and one 3-foot piece. She will have to buy about 15 feet of wood.
7	A	Round each measurement to the nearest whole number, then multiply. 12.2 rounds down to 12 (because of the 2 in the tenths place) and 7.8 rounds up to 8 (because of the 8 in the tenths place). 12 x 8 = 96.
8	B	Round 6.78 to the nearest whole number. Since there is a 7 in the tenths place, round up (to the whole number 7). 7 x 3 = 21.
9	C	The answer options indicate that this problem can be solved using estimation. Round 13.2 down to 13 and 62 down to 60. 13 x 60 = 780, which is closest to the option '800 points.'
10	B	Round each number to the tenths place. 31.245 rounds down to 31.2 and 1.396 rounds up to 1.4. Think of 31.2 - 1.4 as 31.2 - 1.2 (which equals 30) minus another 0.2. That will put the answer slightly less than 30, so it is between 29.5 and 30.

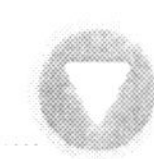

Multiplication of Whole Numbers (5.NBT.B.5)

Question No.	Answer	Detailed Explanation
1	B	79 × 14: 36, 280, 90, + 700 = 1106
2	C	This is a multiplication problem, because it is an array of 18 rows with 50 objects in each row. 50 × 18: 0, 400, 0, + 500 = 900
3	D	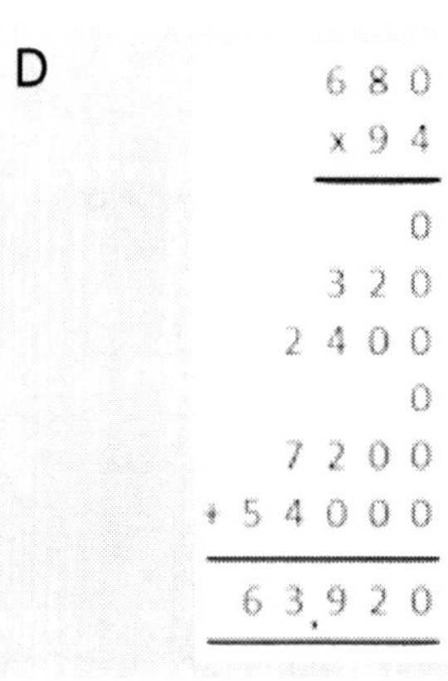
4	C	Multiplying by 11 can be thought of as x 10 and x 1 more. 34 x 10 = 340 and 34 x 1 = 34. Together they equal 374.

Question No.	Answer	Detailed Explanation
5	A	The Commutative Property of Multiplication states that when two factors are multiplied together, the product is the same no matter the order of the factors.
6	D	The array shows three rows with seven objects in each row. There are 21 objects in all. The array is called a 3 by 7 array, which is shown as 3 x 7 = 21.
7	B	According to the Distributive Property of Multiplication, you can break one of the factors (101) into two parts (100 and 1) and multiply them both by the other factor. 596 x 100 and 596 x 1 will produce the same answer as multiplying 596 x 100 and adding 596 more.
8	C	Three of the trays held 15 cookies each, so 3 x 15 = 45. The other six trays held 18 cookies each, so 6 x 18 = 108. To find the total, add 45 + 108 = 153.
9	A	In the first step, 3 x 8 is recorded as 32. It should be 24.
10	B	407 x 35 35 0 2000 210 0 + 12000 14,245

Division of Whole Numbers (5.NBT.B.6)

Question No.	Answer	Detailed Explanation
1	A	The equation 48 ÷ ___ = 12 can be thought of as 48 ÷12 = ___. There are 4 twelves in 48. Check the work by using multiplication (4 x 12 = 48).
2	C	To solve the problem, divide 72 by 8. 72 can be divided evenly by 8. Check the work by using multiplication (8 x 9 = 72).
3	B	1248 divided by 6 is 208 with remainder 0 = 208 R 0 = 208 0/6 **Show Work:** 0 2 0 8 6 | 1 2 4 8 0 1 2 1 2 0 4 0 4 8 4 8 0
4	C	Divide the number of students by the number of seats in each row. 96 ÷ 10 = 9 R 6. The remaining 6 students still had to sit in a row, even though it was not full. The answer is 10 rows.
5	C	6720 divided by 15 is 448 with remainder 0 = 448 R 0 = 448 0/15 **Show Work:** 0 4 4 8 1 5 | 6 7 2 0 0 6 7 6 0 7 2 6 0 1 2 0 1 2 0 0

Name: ______________________ Date: ____________

Question No.	Answer	Detailed Explanation
6	D	To divide by 100, move the decimal point two places to the left.
7	D	There is no number that can be divided by zero.
8	A	100 divided by 12 is 8 with remainder 4 = 8 R 4 = 8 4/12 **Show Work:** 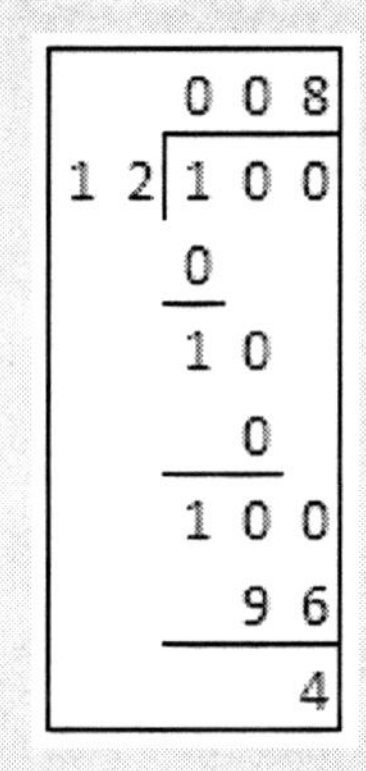 After filling 8 boxes, there will be a remainder of 4 donuts.
9	B	Any number divided by 1 remains the same.
10	B	680 divided by 40 is 17 with remainder 0 = 17 R 0 = 17 0/14 **Show Work:**

```
     0 1 7
4 0 |6 8 0
     0
     -
     6 8
     4 0
     ---
     2 8 0
     2 8 0
    ------
         0
```

Add, Subtract, Multiply, and Divide Decimals (5.NBT.B.7)

Question No.	Answer	Detailed Explanation
1	A	Add each number, maintaining the place value of the digits. Any time a sum exceeds 9, carry the tens to the next highest place value. $$\begin{array}{r} 4.18 \\ 3.75 \\ +\ 3.99 \\ \hline 11.92 \end{array}$$
2	C	Add each number, maintaining the place value of the digits. $$\begin{array}{r} 6.472 \\ 0.01 \\ 3 \\ +\ 0.5 \\ \hline 9.982 \end{array}$$
3	B	Add each number, maintaining the place value of the digits. Any time a sum exceeds 9, carry the tens to the next highest place value. $$\begin{array}{r} 2.09 \\ 2.09 \\ 3.72 \\ +\ 6.60 \\ \hline 14.50 \end{array}$$
4	D	Subtract the numbers, keeping their place values in line. Bring the decimal straight down to the solution. $$\begin{array}{r} 85.37 \\ -\ 75.2 \\ \hline 10.17 \end{array}$$

Question No.	Answer	Detailed Explanation
5	A	Subtract the numbers, keeping their place values in line. Bring the decimal straight down to the solution. 5 14 3.6 4 - 1.4 6 2.1 8
6	B	Subtract the numbers, keeping their place values in line. Bring the decimal straight down to the solution. 0 9 11 12 1 0 2.2 - 9 8.6 0 0 3.6
7	C	To solve, multiply without decimals. Then insert the decimal in your answer. Be sure the product has as many places to the right of the decimal as both factors. \$ 0.4 2 x 8 1 6 + 3 2 0 \$ 3.3 6
8	B	To solve, multiply without decimals. Then insert the decimal in your answer. Be sure the product has as many places to the right of the decimal as both factors. 0.2 5 x 1.1 5 2 0 5 0 2 0 0 0.2 7 5

Question No.	Answer	Detailed Explanation
9	C	To solve, use division. Divide the numbers without the decimal point. Then, insert a decimal into the answer, leaving the same number of places to the right of the decimal as the dividend. 42 ÷ 3 = 14 --> 0.14
10	D	To solve, use division. Move both decimal places to the right one place, so you are dividing by a whole number (0.9 ÷ 3). Divide the numbers without the decimal point. Then, insert a decimal into the answer, leaving the same number of places to the right of the decimal as the dividend (remember that you shifted the decimal to have only one place to the right of the dividend). 9 ÷ 3 = 3 --> 0.3

Numbers and Operations – Fractions

Add & Subtract Fractions (5.NF.A.1)

1. **Add: 2/10 + 1/10 =**

   Ⓐ 3/20
   Ⓑ 3/10
   Ⓒ 1/10
   Ⓓ 2/10

2. **To make a bowl of punch, Joe mixed 1 1/4 gallons of juice with 1 2/4 gallons of sparkling water. How much punch does he have?**

   Ⓐ 2 3/4 gallons
   Ⓑ 3 gallons
   Ⓒ 1/4 gallon
   Ⓓ 3/4 gallon

3. **Subtract: 3/4 - 2/4 =**

   Ⓐ 5/4
   Ⓑ 1/4
   Ⓒ 3/4
   Ⓓ 1

4. **Subtract: 3 4/10 - 1 1/10 =**

   Ⓐ 1 3/10
   Ⓑ 2 1/10
   Ⓒ 3 3/10
   Ⓓ 2 3/10

5. **To add the fractions 3/4 and 7/12, what must first be done?**

   Ⓐ Reduce the fractions to lowest terms
   Ⓑ Change to improper fractions
   Ⓒ Make the numerators the same
   Ⓓ Find a common denominator

6. **Add: 1/2 + 1/4 =**

   Ⓐ 2/6
   Ⓑ 2/3
   Ⓒ 3/4
   Ⓓ 1/2

7. **Find the difference: 2/3 - 1/9 =**

   Ⓐ 1/6
   Ⓑ 5/9
   Ⓒ 3/12
   Ⓓ 2/27

8. **Find the sum: 2 1/8 + 5 1/2 =**

   Ⓐ 7 2/10
   Ⓑ 10 1/16
   Ⓒ 3 1/6
   Ⓓ 7 5/8

9. **Find the sum of five and five eighths plus one and one fourth.**

   Ⓐ 6 7/8
   Ⓑ 10 6/8
   Ⓒ 6 6/12
   Ⓓ 7 2/10

10. **Subtract: 5 - 1/3 =**

    Ⓐ 5 1/3
    Ⓑ 4 1/3
    Ⓒ 3 2/3
    Ⓓ 4 2/3

11. **Jordan had a plank of wood that was 8 5/16 inches long. He sawed off 2 3/16 inches. Now how long is the plank of wood?**

    Ⓐ 10 8/32 inches
    Ⓑ 6 1/4 inches
    Ⓒ 6 2/16 inches
    Ⓓ 10 8/16 inches

12. At the beginning of 5th grade, Amber's hair was 8 1/2 inches long. By the end of 5th grade it was 10 3/4 inches long. How many inches did Amber's hair grow during 5th grade?

Ⓐ 19 1/4 inches
Ⓑ 18 4/6 inches
Ⓒ 2 1/2 inches
Ⓓ 2 1/4 inches

13. Solve: 1/5 + 3/5 + 4/5 = ___________

Ⓐ 1 3/5
Ⓑ 8/15
Ⓒ 7/5
Ⓓ 5/8

14. Find the missing number: 4 1/4 + __________ = 7 1/2

Ⓐ 3 1/4
Ⓑ 3 3/4
Ⓒ 3 1/2
Ⓓ 2 3/4

15. Solve: 7/10 - (4/10 - 1/10) =

Ⓐ 2/10
Ⓑ 4/10
Ⓒ 0
Ⓓ 3/10

Name: ____________________ Date: ____________

Problem Solving with Fractions (5.NF.A.2)

1. Susan's homework was to practice the piano for 3/4 of an hour each night. How many minutes each night did she practice?

   Ⓐ 30 minutes
   Ⓑ 15 minutes
   Ⓒ 45 minutes
   Ⓓ 60 minutes

2. Three fifths of the 30 students are boys. How many students are girls?

   Ⓐ 12 girls
   Ⓑ 18 girls
   Ⓒ 6 girls
   Ⓓ 8 girls

3. Walking at a steady pace, Ella walked 11 miles in 3 hours. Which mixed number shows how many miles she walked in an hour?

   Ⓐ 2/3
   Ⓑ 2 2/3
   Ⓒ 3
   Ⓓ 3 2/3

4. Arthur had one dollar. He spent 75 cents of that dollar. What fraction of his whole dollar did he spend?

   Ⓐ 1/2
   Ⓑ 3/4
   Ⓒ 1/4
   Ⓓ 2/4

5. In science class we discovered that 7/8 of an apple is water. What fraction of the apple is not water?

   Ⓐ 1/6
   Ⓑ 1/7
   Ⓒ 7/8
   Ⓓ 1/8

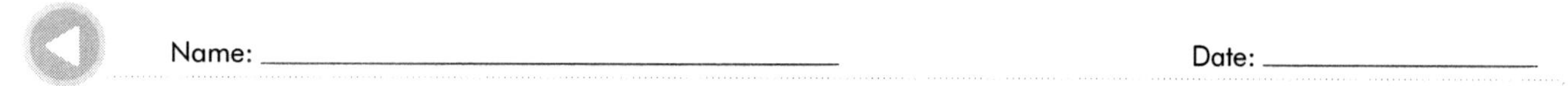
Name: ______________________ Date: ______________

6. There were 20 pumpkins in a garden. One fourth of the pumpkins were too small, one tenth were too large, and one half were just the right size. The rest were not ripe yet. How many of the pumpkins were too small?

Ⓐ 3
Ⓑ 2
Ⓒ 5
Ⓓ 10

7. Timothy decided to clean out his closet by donating some of his 45 button-down shirts. He gave away 9 shirts. What fraction of the shirts did he give away?

Ⓐ 1/5
Ⓑ 1/9
Ⓒ 1/2
Ⓓ 36/45

8. Lindsey wants a new bike that costs $230. Her father said that if she saves up 30% (or 30/100) of the cost, he will pay the rest. How much money does Lindsey need to save?

Ⓐ $23
Ⓑ $30
Ⓒ $69
Ⓓ $72

9. There are 32 students in Mr. Duffy's class. If 4 come to after school tutoring, what fraction of the class comes to after school tutoring?

Ⓐ 28/32
Ⓑ 2/16
Ⓒ 1/4
Ⓓ 2/8

10. Mike is buying a pair of jeans. They normally cost $45 but they are on sale for 20% off. How much will Mike pay for the jeans (not including tax)?

Ⓐ $9
Ⓑ $43
Ⓒ $25
Ⓓ $36

11. In a recital, there were 15 boys dancing. If 1/3 of the dancers were boys, how many dancers were there in all?

Ⓐ 45
Ⓑ 50
Ⓒ 30
Ⓓ 60

12. Dara has to solve 35 math problems for homework. She has completed 14 of them. What fraction of the problems does she have left to do?

Ⓐ 14/35
Ⓑ 3/5
Ⓒ 14/21
Ⓓ 2/5

13. A 5th grade volleyball team scored 32 points in one game. Of those points, 2/8 were scored in the second half. How many points were scored in the first half of the game?

Ⓐ 12
Ⓑ 4
Ⓒ 20
Ⓓ 24

14. A recipe to make 48 cookies calls for 3 cups of flour. However, you do not want to make 48 cookies, but only 24 cookies. Which fraction shows how much flour to use?

Ⓐ 2 cups
Ⓑ 1 2/3 cups
Ⓒ 1 1/2 cups
Ⓓ 2 2/3 cups

15. Coleen had a choice between 3/10 of a bag of candy or 3/6 of a bag of candy. If she wanted to get as much candy as possible, which one should she choose?

Ⓐ 3/10 because the numerator is larger
Ⓑ 3/10 because the denominator is larger
Ⓒ 3/6 because the denominator is smaller
Ⓓ 3/6 because the numerator is smaller

Interpreting Fractions (5.NF.B.3)

1. **Suppose three friends wanted to share four cookies equally. How many cookies would each friend receive?**

   Ⓐ 1 1/3
   Ⓑ 3/4
   Ⓒ 1 3/4
   Ⓓ 1/3

2. **If 18 is divided by 5, which fraction represents the remainder?**

   Ⓐ 3/18
   Ⓑ 3/5
   Ⓒ 5/10
   Ⓓ 1/3

3. **If there are 90 minutes in a soccer game and 4 squads of players will share this time equally, how many minutes will each squad play?**

   Ⓐ 22/4
   Ⓑ 22 1/2
   Ⓒ 22 2/10
   Ⓓ 16 4/22

4. **Damien has $695 in the bank. He wants to take out 2/5 of his money. If he uses a calculator to figure out this amount, which buttons should he press?**

   Ⓐ [6] [9] [5] [x] [2] [x] [5] [=]
   Ⓑ [6] [9] [5] [÷] [2] [x] [5] [=]
   Ⓒ [6] [9] [5] [÷] [2] [÷] [5] [=]
   Ⓓ [6] [9] [5] [x] [2] [÷] [5] [=]

5. **Five friends are taking a trip in a car. They want to share the driving equally. If the trip takes 7 hours, how long should each friend drive?**

   Ⓐ 5/7 of an hour
   Ⓑ 1 hour 7 minutes
   Ⓒ 1 2/5 hours
   Ⓓ 1 hour 2 minutes

Name: ________________________ Date: ____________

Multiply Fractions (5.NF.B.4.A)

1. **Multiply: 2/3 x 4/5 =**

   Ⓐ 8/15
   Ⓑ 3/4
   Ⓒ 6/8
   Ⓓ 4/15

2. **Find the product: 5 x 2/3 x 1/2 =**

   Ⓐ 1 1/3
   Ⓑ 5
   Ⓒ 2 2/3
   Ⓓ 1 2/3

3. **Which of the following is equivalent to 5/6 x 7?**

   Ⓐ 5 ÷ (6 x 7)
   Ⓑ (5 x 7) ÷ 6
   Ⓒ (6 x 7) ÷ 5
   Ⓓ (1 ÷ 7) x (5 ÷ 6)

4. **Which of the following is equivalent to 4/10 x 3/8?**

   Ⓐ 4 ÷ (10 x 3) ÷ 8
   Ⓑ (4 + 3) x (10 + 8)
   Ⓒ (4 x 3) ÷ (10 x 8)
   Ⓓ (4 - 3) ÷ (10 - 8)

5. **Hector is using wood to build a dog house. Each wall is 4/7 of a yard tall and 3/5 of a yard wide. Knowing that the area of each wall is the base times the height, how many square yards of wood will he need to build 4 walls of equal size?**

   Ⓐ 1 2/3
   Ⓑ 1 13/35
   Ⓒ 12/35
   Ⓓ 1 4/12

Multiply to Find Area (5.NF.B.4.B)

1. Dominique is covering the top of her desk with contact paper. The surface measures 7/8 yard by ¾ yard. How much contact paper will she need to cover the surface of the desk top?

Ⓐ 21/32 yd^2
Ⓑ 13/8 yd^2
Ⓒ 20/24 yd^2
Ⓓ 1 5/8 yd^2

2. Christopher is tiling his bathroom floor with tiles that are each 1 square foot. The floor measures 2 ½ feet by 3 ¾ feet. How many tiles will he need to cover the floor?

Ⓐ 6 3/8
Ⓑ 6 ¼
Ⓒ 9 3/8
Ⓓ 8

3. Lin and Tyra are measuring the area of the piece of paper shown below. Lin multiplied the length times the width to find an answer. Tyra traced the paper onto 1-inch graph paper and counted the number of squares. How should their answers compare?

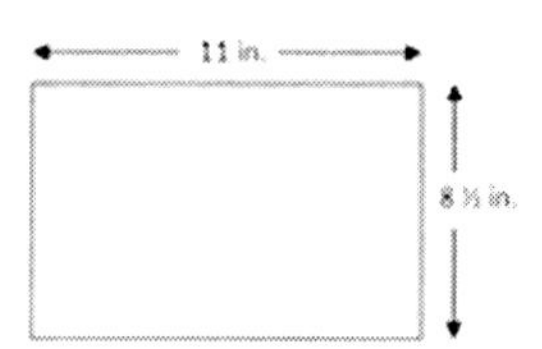

Ⓐ Lin's answer will be a mixed number, but Tyra's will be a whole number.
Ⓑ Tyra's answer will be greater than Lin's answer.
Ⓒ Lin's answer will be greater than Tyra's answer.
Ⓓ They should end up with almost exactly the same answer.

4. Jeremy found that it takes 14 centimeter cubes to cover the surface of a rectangular image. Which of these measurements could possibly be the length and width of the rectangle he covered?

Ⓐ Length = 3 ½ cm, width = 4 cm
Ⓑ Length = 4 ½ cm, width = 3 cm
Ⓒ Length = 7 cm, width = 7 cm
Ⓓ Length = 5 ½ cm, width = 8 ½ cm

5. What is the area of the court shown below?

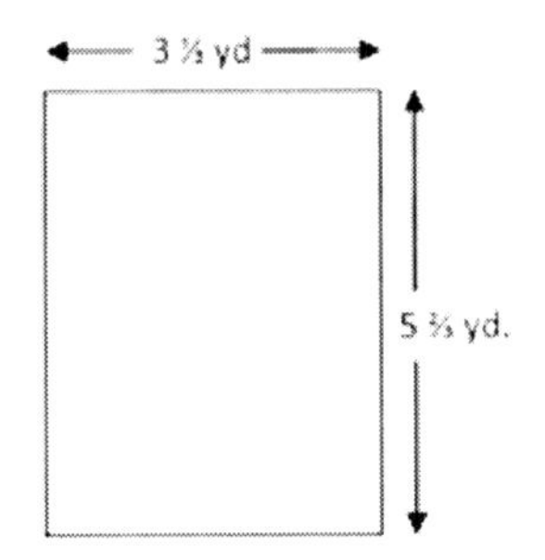

Ⓐ **15 2/9 yd²**
Ⓑ **18 8/9 yd²**
Ⓒ **16 yd²**
Ⓓ **9 yd²**

Multiplication as Scaling (5.NF.B.5.A)

1. If y is four times as much as z, which number completes this equation?
   z * ___ = y

   Ⓐ 4
   Ⓑ 0.4
   Ⓒ 1/4
   Ⓓ 40

2. In the equation a * b = c, if b is a fraction greater than 1, then c will be _______.

   Ⓐ a mixed number
   Ⓑ less than b
   Ⓒ greater than a
   Ⓓ equal to b ÷10

3. If d * e = f, and e is a fraction less than 1, then f will be _______.

   Ⓐ greater than d
   Ⓑ less than d
   Ⓒ equal to e ÷ d
   Ⓓ less than 1

4. In which equation is r less than s?

   Ⓐ r - 6 = s
   Ⓑ s * 6 = r
   Ⓒ r ÷ 6 = s
   Ⓓ s * 1/6 = r

5. Ryan and Alex are using beads to make necklaces. Ryan used one fifth as many beads as Alex. Which equation is true?

   Ⓐ R * 1/5 = A
   Ⓑ R = A * 5
   Ⓒ R * 5 = A
   Ⓓ R ÷ 5 = A

Name: ______________________ Date: ______________

Numbers Multiplied by Fractions (5.NF.B.5.B)

1. **Which statement is true about the following equation?**
   6, 827 x 2/7 = ?

   Ⓐ The product will be less than 6,827.
   Ⓑ The product will be greater than 6,827.
   Ⓒ The product will be less than 2/7.
   Ⓓ The product will be equal to 6,827 ÷ 7.

2. **Which statement is true about the following equation?**
   27,093 x 5/4 = ?

   Ⓐ The product will be equal to 27,093 ÷ 54.
   Ⓑ The product will be less than 5/4.
   Ⓒ The product will be less than 27,093.
   Ⓓ The product will be greater than 27,093.

3. **Which is correct answer?**
   18,612 x 1 1/7 = ?

   Ⓐ 15,000
   Ⓑ 21,000
   Ⓒ 38,000
   Ⓓ 2,500

4. **Which number completes the equation?**
   3,606 x ___ = 4,808

   Ⓐ 2/3
   Ⓑ 4/3
   Ⓒ 2 ½
   Ⓓ 9/3

5. **Which number completes the equation?**
   ___ x 5/6 = 17,365

   Ⓐ 5,838
   Ⓑ 50,838
   Ⓒ 20,838
   Ⓓ 10,838

Real World Problems with Fractions (5.NF.B.6)

1. Chef Chris is using ¾ lb. of chicken per person at a luncheon. If there are 17 people at the luncheon, how many pounds of chicken will he use?

   Ⓐ 12 3/4
   Ⓑ 51/68
   Ⓒ 48/4
   Ⓓ 17 3/4

2. A team of runners ran a relay race 9/10 of a mile long. If Carl ran 3/5 of the race, how far did his teammates run?

   Ⓐ 9/25 mile
   Ⓑ 27/50 mile
   Ⓒ 1/10 mile
   Ⓓ 2/5 mile

3. There are 1 4/5 lbs. of jelly beans in each bag. If Mrs. Lancer buys 3 bags of jelly beans for her class, how many pounds of jelly beans will she have in all?

   Ⓐ 3 12/15
   Ⓑ 5 2/5
   Ⓒ 3 4/15
   Ⓓ 5 4/5

4. Mario is in a bike race that is 3 1/5 miles long. He gets a flat tire 2/3 of the way into the race. How many miles did he make it before he got a flat tire?

   Ⓐ 3 2/15
   Ⓑ 1 3/8
   Ⓒ 2 2/15
   Ⓓ 2/3

5. Jackson is swimming laps in a pool that is 20½ meters long. He swims 4½ laps. How many meters did he swim?

   Ⓐ 80 1/4
   Ⓑ 92 1/4
   Ⓒ 84 1/2
   Ⓓ 90

Dividing Fractions (5.NF.B.7.A)

1. Divide: $2 \div 1/3 =$

Ⓐ 3
Ⓑ 2
Ⓒ 1
Ⓓ 6

2. In order to divide by a fraction you must first:

Ⓐ find its reciprocal
Ⓑ match its denominator
Ⓒ find its factors
Ⓓ multiply by the numerator

3. Divide: $3 \div 2/3 =$

Ⓐ 4 2/3
Ⓑ 3 2/3
Ⓒ 4
Ⓓ 4 1/2

4. Complete the following:
Dividing a number by a fraction less than 1 results in a quotient that is ______________________ the original number.

Ⓐ the reciprocal of
Ⓑ less than
Ⓒ greater than
Ⓓ equal to

5. 5 people want to evenly share a 1/3 pound bag of peanuts. How many pounds should each person get?

Ⓐ 3/5
Ⓑ 1 2/3
Ⓒ 3/15
Ⓓ 1/15

Name: ______________________ Date: ______________

Dividing by Unit Fractions (5.NF.B.7.B)

1. Which best explains why $6 \div \frac{1}{4} = 24$?

Ⓐ $24 \div \frac{1}{4} = 6$
Ⓑ $24 \times \frac{1}{4} = 6$
Ⓒ $24 \div 6 = \frac{1}{4}$
Ⓓ $24 = \frac{1}{4} \times 6$

2. Which model best represents the following equation?
$4 \div 1/3 = 12$

Ⓐ

Ⓑ

Ⓒ

Ⓓ

3. Which equation matches this model?

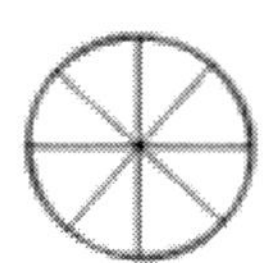 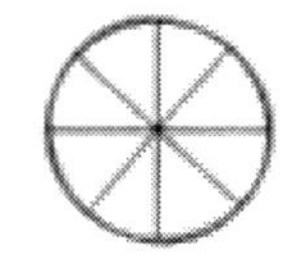 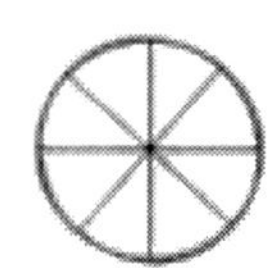

Ⓐ $24 \div 1/8 = 3$
Ⓑ $24 \div 1/3 = 8$
Ⓒ $8 \div 1/3 = 24$
Ⓓ $3 \div 1/8 = 24$

4. Byron has 5 pieces of wood from which to build his birdhouse. If he cuts each piece into fifths, how many pieces will he have?

Ⓐ 25
Ⓑ 5
Ⓒ 1/5
Ⓓ 5/25

5. Angelina has 10 yards of fabric. She needs 1/3 yard of fabric for each purse she will sew. How many purses will she be able to make?

Ⓐ **3 1/3**
Ⓑ **10 1/3**
Ⓒ **30**
Ⓓ **13**

Name: ______________________ Date: ______________

Real World Problems Dividing Fractions (5.NF.B.7.C)

1. **Darren has a 3 cup bag of snack mix. Each serving is ¼ cup. Which model will help him determine how many ¼ cup servings are in the whole bag of snack mix?**

Ⓐ 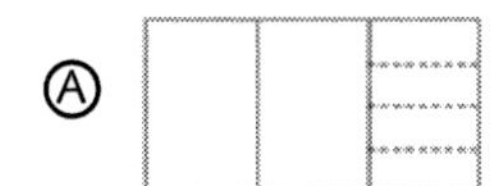

Ⓑ 

Ⓒ 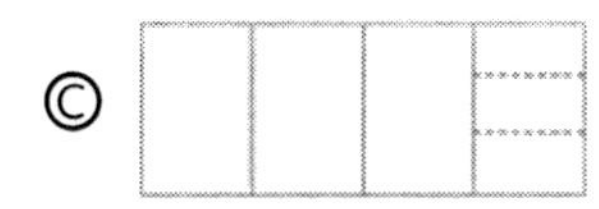

Ⓓ 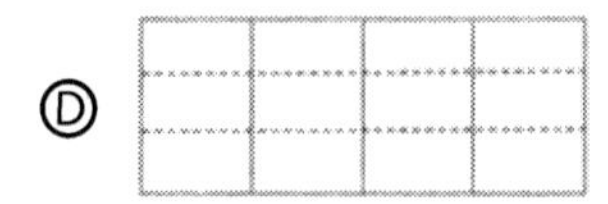

2. **Which situation could be represented by the following model?**

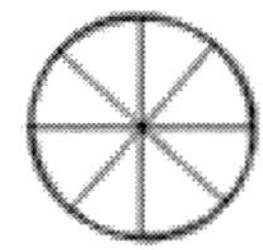 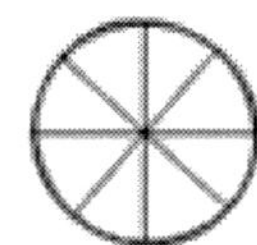  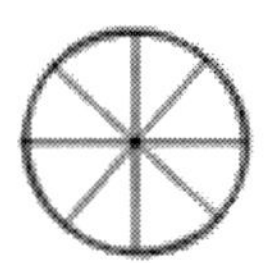

Ⓐ The number of 1/8 lb. servings of cheese in 4 lbs. of cheese
Ⓑ Four friends share 8 lbs. of cheese
Ⓒ 4 lbs. of cheese divided into 8 equal servings
Ⓓ The amount of cheese needed for 8 people to each have ¼ lb.

3. **A team of 3 runners competes in a 1/4 mile relay race. If each person runs an equal portion of the race, how far does each person run?**

Ⓐ 3/7 mile
Ⓑ 3/12 mile
Ⓒ ¾ mile
Ⓓ 1/12 mile

4. **A beaker holds 1/10 of a liter of water. If the water is divided equally into 6 test tubes, how much water will be in each test tube?**

Ⓐ 1/60 liter
Ⓑ 6/10 liter
Ⓒ 1/6 liter
Ⓓ 10/16 liter

5. Mrs. Blake orders 3 pizzas for a school party. If each slice is 1/12 of a pizza, how many slices are there in all?

Ⓐ 24
Ⓑ 4 1/3
Ⓒ 36
Ⓓ 3/12

End of Numbers and Operations – Fractions

Numbers and Operations – Fractions

Answer Key
&
Detailed Explanations

Add & Subtract Fractions (5.NF.A.1)

Question No.	Answer	Detailed Explanation
1	B	When fractions have a common denominator (in this case 10), just add the numerators (2 + 1 = 3) and keep the denominator the same.
2	A	Add the whole numbers (1+1) to get 2. Then add the fractions. As they have a common denominator of 4, just add the numerators (1+2) to get 3/4. The total is 2 3/4.
3	B	As the fractions have a common denominator of 4, just subtract the numerators (3 - 2) to get 1/4.
4	D	Subtract the whole numbers (3 - 1) to get 2. Then subtract the fractions. Since they have a common denominator of 10, just subtract the numerators (4 - 1) to get 3/10. The total is 2 3/10.
5	D	Fractions must have a common denominator to be added. Multiply 3/4 x 3 to get 9/12 so that both terms have a denominator of 12.
6	C	Fractions must have a common denominator to be added. Multiply 1/2 x 2 to get 2/4. Then add the numerators (2 + 1) and keep the denominator the same to get 3/4.
7	B	Fractions must have a common denominator to be subtracted. Multiply 2/3 x 3 to get 6/9. Then subtract the numerators (6 - 1) and keep the denominator the same to get 5/9.
8	D	First add the whole numbers (2 + 5) to get 7. Then add the fractions. Since fractions must have a common denominator to be added, multiply 1/2 x 4 to get 4/8. Then add the numerators (1+4) to get 5/8. The total is 7 5/8.
9	A	First add the whole numbers (5 + 1) to get 6. Then add the fractions (5/8 + 1/4). Since fractions must have a common denominator to be added, multiply 1/4 x 2 to get 2/8. Then add the numerators (5 + 2) to get 7/8. The total is 6 7/8.

Question No.	Answer	Detailed Explanation
10	D	In order to subtract a fraction from a whole number, convert 1 from the whole number into a fraction with a common denominator. The number 1 can be converted to thirds by changing it to 3/3. That leaves 4 3/3 - 1/3. Keep 4 as the whole number and subtract the numerators of the fractions to get 2/3.
11	C	Subtract the whole numbers (8 - 2) to get 6. Then subtract the fractions. Since they have a common denominator of 16, just subtract the numerators (5 - 3) to get 2/16. The total is 6 2/16.
12	D	To find the difference, subtract the starting length from the ending length. First subtract the whole numbers (10 - 8) to get 2. Then subtract the fractions. As they do not have a common denominator, multiply 1/2 x 2 to get 2/4. Then just subtract the numerators (3 - 2) to get 1/4. The difference is 2 1/4.
13	A	Since the fractions all have a common denominator (5), just add the numerators. 1 + 3 + 4 = 8. The total is 8/5 which is equivalent to 1 3/5 (because 8/5 is really 5/5 and 3/5).
14	A	As this is a missing addend problem, it can be solved by subtracting 7 1/2 - 4 1/4. First subtract the whole numbers (7 - 4) to get 3. Then subtract the fractions. Since they do not have a common denominator, multiply 1/2 x 2 to get 2/4. Then, just subtract the numerators (2 - 1) to get 1/4. The total is 3 1/4.
15	B	First, complete the part of the problem in parentheses. As the fractions have a common denominator (10), just subtract the numerators to get 3/10. Then subtract the 3/10 from 7/10 to get 4/10.

Problem Solving with Fractions (5.NF.A.2)

Question No.	Answer	Detailed Explanation
1	C	Multiply 60 (the number of minutes in an hour) by 3/4 to find the number of minutes she practiced. $60 \times \frac{3}{4} = \frac{180}{4}$ $= 45$
2	A	Multiply 30 by 3/5 to find the number of boys. $30 \times \frac{3}{5} = \frac{90}{5}$ $= 18$ If there are 18 boys, there must be 12 girls (30 - 18 = 12).
3	D	To solve, divide 11 miles by 3 hours. Convert the improper fraction to a mixed number. $\frac{11}{3} = 3\frac{2}{3}$
4	B	Arthur spent 75/100. In lowest terms (divided by 25), this number is ¾.
5	D	A whole apple is 1, or 8/8. To find out how much is not water, subtract the fraction that is water. $\frac{8}{8} - \frac{7}{8} = \frac{1}{8}$
6	C	This problem has a lot of extra information. To find the number of pumpkins that were too small, just multiply the total number of pumpkins (20) by the fraction of pumpkins that were too small (1/4). $20 \times \frac{1}{4} = \frac{20}{4}$ $= 5$ Convert from an improper fraction to a whole number by dividing 20 by 4.

Question No.	Answer	Detailed Explanation
7	A	Timothy gave away 9 out of 45 shirts. This is the fraction 9/45. Since that option is not available, reduce the fraction to lowest terms by dividing: $\frac{9}{45} \div \frac{9}{9} = \frac{1}{5}$
8	C	30% is the fraction 30/100 or 3/10. Multiply the price of the bike (230) by Lindsey's share (3/10): $230 \times \frac{3}{10} = \frac{690}{10} = 69$ Convert from an improper fraction to a whole number by dividing 690 by 10.
9	B	4 out of 32 students come to tutoring. This is the fraction 4/32. Since that option is not available, reduce the fraction by dividing: $\frac{4}{32} \div \frac{2}{2} = \frac{2}{16}$
10	D	20% is the fraction 20/100 or 2/10. Multiply the price of the jeans (45) by the sale discount (2/10): $\frac{45}{1} \times \frac{2}{10} = \frac{90}{10} = 9$ Since this \$9 represents Mike's savings, he will have to pay \$36 (45 - 9 = 36).
11	A	If 1/3 = 15, then 3/3 (all ofthe dancers) would equal 3 x 15, which is 45.
12	B	She has completed 14/35 problems. This means she has 21 left to do (35 - 14 = 21). 21 out of 35 is the fraction 21/35. Since that option is not available, reduce the fraction by dividing: $\frac{21}{35} \div \frac{7}{7} = \frac{3}{5}$

Question No.	Answer	Detailed Explanation
13	D	Multiply 32 by 2/8 to find the number of points they scored in the second half. $32 \times \frac{2}{8} = \frac{64}{8}$ $= 8$ If they scored 8 points in the second half, they must have scored 24 points in the first half (32 - 8 = 24).
14	C	You will only need half the amount of flour, since 24 is half of 48. Multiply 3 cups by 1/2 to find out how much flour to use. $3 \times \frac{1}{2} = \frac{3}{2}$ $= 1\frac{1}{2}$
15	C	Since the numerators are the same, compare the denominators. Tenths are smaller portions than sixths, so three small portions is less than three larger portions. Three sixths is the larger amount.

Interpreting Fractions (5.NF.B.3)

Question No.	Answer	Detailed Explanation
1	A	The first three cookies can be shared by having each friend receive 1 whole cookie. That leaves 1 cookie to be divided among the three friends. This can be shown as a fraction with the dividend (1) as the numerator and the divisor (3) as the denominator. Each friend will receive 1 whole cookie and 1/3 of the last cookie that was divided.
2	B	The number 5 goes into 18 three whole times (5 x 3 = 15), leaving a remainder of 3. That three can be divided by 5 by using the fraction 3/5.
3	B	To solve, divide 90 minutes by 4 squads. This creates the improper fraction 90/4. To change it to a mixed number, divide 90 by 4 to get 22 remainder 2. The remainder of 2 also needs to be divided among the 4 squads, so it becomes the fraction 2/4, or 1/2. Each squad will play for 22 1/2 minutes.
4	D	To find 2/5 of 695, multiply the whole number by the fraction. Since 2/5 is really 2 ÷ 5, this means you will multiply 695 x 2 ÷ 5.

Question No.	Answer	Detailed Explanation
5	C	7 hours divided by 5 people is the fraction 7/5. Of this, 5/5 equals one whole, leaving 2/5 as a fraction. These 2/5 are not 2 minutes, they are a fraction of an hour. The total time is 1 2/5.

Multiply Fractions (5.NF.B.4.A)

Question No.	Answer	Detailed Explanation
1	A	First, multiply the numerators (2 x 4 = 8) then multiply the denominators (3 x 5 = 15) to get the fraction 8/15.
2	D	Multiply the first two terms first, using 5/1 for the whole number 5. 5/1 x 2/3 = 10/3. Then multiply this fraction by the third term: 10/3 x 1/2 = 10/6 Change the improper fraction 10/6 to a mixed number by dividing 10 by 6. Then change 1 4/6 into lowest terms, which is 1 2/3.
3	B	Multiplying a fraction by a whole number is the same as multiplying the numerator by a whole number then dividing the product by the denominator.
4	C	The product of two fractions is equal to the product of the numerators divided by the product of the denominators.
5	B	To solve, multiply 4/7 x 3/5 x 4. Multiply the first two terms first: 4/7 x 3/5 = 12/35 Then multiply this fraction by 4. Remember that the whole number 4 can be shown as the fraction 4/1. 12/35 x 4 = 48/35 Since 35/35 is 1 whole the fraction can be shown as the mixed number 1 13/35.

Multiply to Find Area (5.NF.B.4.B)

Question No.	Answer	Detailed Explanation
1	A	Find the area of the desk top by multiplying: 7/8 yd x ¾ yd = 21/32 yd^2
2	C	Find the area of the floor by multiplying: 2 ½ x 3 ¾ = 5/2 x 15/4 = 75/8 = 9 3/8

Question No.	Answer	Detailed Explanation
3	D	Multiplying length x width of a rectangle and tiling the rectangle with unit squares are both accurate ways to determine area. Therefore, Lin and Tyra should both end up with the same answer, or nearly the same answer (since counting fractional parts of tiles isn't as precise as multiplying).
4	A	Multiplying length x width of a rectangle should produce the same number as tiling the rectangle with unit squares. Therefore, multiply to find that: 3 ½ x 4 = 7/2 x 4/1 = 28/2 = 14
5	B	Find the area of the court by multiplying: 3 1/3 x 5 2/3 = 10/3 x 17/3 = 170/9 = 18 8/9 yd^2

Multiplication as Scaling (5.NF.B.5.A)

Question No.	Answer	Detailed Explanation
1	A	When multiplying two numbers (a and b) greater than 1, the product will be 'a' times as much as 'b' or 'b' times as much as 'a'.
2	C	When multiplying, if one factor is a fraction greater than 1, the product will be greater than the other factor.
3	B	When multiplying, if one factor is a fraction less than 1, the product will be less than the other factor.
4	D	When multiplying, if one factor is a fraction less than 1, the product will be less than the other factor.
5	C	If Ryan has 1/5 as many beads as Alex, then Alex has five times as many beads as Ryan. The way to show this is by multiplying Ryan's beads by 5 to equal Alex's beads.

Numbers Multiplied by Fractions (5.NF.B.5.B)

Question No.	Answer	Detailed Explanation
1	A	Multiplying a number by a fraction less than 1 will result in a product that is less than the original number.
2	D	Multiplying a number by a fraction greater than 1 will result in a product that is greater than the original number.
3	B	Multiplying a number by a fraction greater than 1 will result in a product that is greater than the original number. Since the second factor is only 1/7 more than one, the product will be just slightly greater than 18,612. The only other option that is greater than 18,612 (option C: 36,000) is more than twice the original number.
4	B	Multiplying a number by a fraction greater than 1 will result in a product that is greater than the original number. Since the product is only slightly greater than the original number, the other factor will be just slightly greater than 1. Therefore, 4/3 (which is equal to 1 1/3) is the only option possible.
5	C	Multiplying a number by a fraction less than 1 will result in a product that is less than the original number. Since the fraction is only slightly less than 1, the other factor will be just slightly greater than 17,365. Therefore, 20,838 is the only option possible, as 50,838 is more than double the product.

Real World Problems with Fractions (5.NF.B.6)

Question No.	Answer	Detailed Explanation
1	A	To multiply a whole number by a fraction, represent the whole number as 17/1. Then, multiply numerators (17 x 3 = 51) to find the numerator and multiply denominators (1 x 4 = 4) to find the denominator. Change the improper fraction 51/4 to a mixed number by dividing 51 by 4 to equal 12 3/4.
2	A	To find how far the teammates ran, subtract 3/5 (Carl's distance) from 5/5 (the total distance) to get 2/5. Then, multiply this fraction by the distance of the race. Multiply numerators (2 x 9 = 18) to find the numerator and multiply denominators (5 x 10 = 50) to find the denominator. Reduce the fraction 18/50 to 9/25.

Question No.	Answer	Detailed Explanation
3	B	To multiply a whole number by a mixed number, first change the whole number to a fraction (3/1) and change the mixed number to a fraction (9/5). Multiply numerators (3 x 9 = 27) to find the numerator and multiply denominators (1 x 5 = 5) to find the denominator. The improper fraction 27/5 can be changed to the mixed number 5 2/5.
4	C	To multiply a fraction by a mixed number, change the mixed number to a fraction (16/5). Multiply numerators (2 x 16 = 32) to find the numerator and multiply denominators (3 x 5 = 15) to find the denominator. The improper fraction 32/15 can be changed to the mixed number 2 2/15.
5	B	To multiply a mixed number by a mixed number, change each mixed number to a fraction (41/2 and 9/2). Multiply numerators (41 x 9 = 369) to find the numerator and multiply denominators (2 x 2 = 4) to find the denominator. The improper fraction 369/4 can be changed to the mixed number 92 1/4.

Dividing Fractions (5.NF.B.7.A)

Question No.	Answer	Detailed Explanation
1	D	The first step in dividing by a fraction is to find its reciprocal, which is the reverse of its numerator and denominator. The fraction 1/3 becomes 3/1, or the whole number 3. Then solve by multiplying. 2x3=6.
2	A	The first step in dividing by a fraction is to find its reciprocal, which is the reverse of its numerator and denominator.
3	D	The first step in dividing by a fraction is to find its reciprocal, which is the reverse of its numerator and denominator. The fraction 2/3 becomes 3/2. Then solve by multiplying (use 3/1 for the whole number 3): 3/1 X 3/2 = 9/2 = 4 1/2.
4	C	Dividing by a number less than one causes the original number to become larger. When dividing by a fraction, multiplying by its reciprocal will create a situation in which you multiply by a larger number and divide by a smaller number, therefore increasing the size.
5	D	To divide 1/3 by 5, multiply 1/3 by the reciprocal of 5, which is 1/5. 1/3 X 1/5 = 1/15.

Dividing by Unit Fractions (5.NF.B.7.B)

Question No.	Answer	Detailed Explanation
1	B	Division can be checked by multiplying the quotient by the divisor to equal the dividend. In this case, 24 x ¼ = 24/4 = 6.
2	D	In option D, each of four units is divided into thirds, resulting in a total of 12 units. Option B also produces 12 units, but it shows 3 units divided into fourths.
3	D	The model shows each of three units divided into eighths, resulting in a total of 24 units. That is shown as 3 ÷ 1/8 = 24. Although option C is a true statement, it does not represent the model.
4	A	To solve, divide the 5 pieces of wood into fifths: 5 ÷ 1/5 = 5 x 5 = 25
5	C	To solve, divide the 10 yards of fabric into thirds: 10 ÷ 1/3 = 10 x 3 = 30

Real World Problems Dividing Fractions (5.NF.B.7.C)

Question No.	Answer	Detailed Explanation
1	B	In option B, each of 3 units is divided into fourths, resulting in a total of 12 units. Option D also produces 12 units, but it shows 4 units divided into thirds.
2	A	The model shows each of 4 units divided into eighths, resulting in a total of 32 units. That shows how many 1/8 units there are in the 4 whole units.
3	D	Divide ¼ mile by 3 to solve: ¼ ÷ 3 = ¼ x 1/3 = 1/12
4	A	Divide 1/10 liter by 6 to solve: 1/10 ÷ 6 = 1/10 x 1/6 = 1/60
5	C	To solve, divide the 3 pizzas by 1/12: 3 ÷ 1/12 = 3 x 12 = 36

Name: ______________________ Date: ______________

Measurement and Data

Converting Units of Measure (5.MD.A.1)

1. **Solve.**
   1 foot = ____

   Ⓐ 1 yard
   Ⓑ 100 centimeters
   Ⓒ 12 inches
   Ⓓ 10 inches

2. **Solve.**
   1 pound = ____

   Ⓐ 12 ounces
   Ⓑ 16 ounces
   Ⓒ 100 grams
   Ⓓ 1 kilogram

3. **Solve.**
   100 centimeters = ____

   Ⓐ 1 foot
   Ⓑ 10 inches
   Ⓒ 1 kilometer
   Ⓓ 1 meter

4. **Solve.**
   1 gallon = ____

   Ⓐ 4 quarts
   Ⓑ 1 liter
   Ⓒ ½ quart
   Ⓓ 4 cups

5. **Solve.**
   1 yard = ____

   Ⓐ 1 meter
   Ⓑ 3 feet
   Ⓒ 12 feet
   Ⓓ 100 inches

6. **Solve.**
   1 kilogram = ____

   Ⓐ **100 grams**
   Ⓑ **1 pound**
   Ⓒ **1,000 grams**
   Ⓓ **100 ounces**

7. **Solve.**
   1 hour = ____

   Ⓐ **3,600 seconds**
   Ⓑ **60 seconds**
   Ⓒ **100 minutes**
   Ⓓ **360 minutes**

8. **Solve.**
   1 liter = ____

   Ⓐ **1,000 centiliters**
   Ⓑ **4 cups**
   Ⓒ **1 gallon**
   Ⓓ **1,000 milliliters**

9. **Solve.**
   1 mile = ____

   Ⓐ **1,000 yards**
   Ⓑ **1 kilometer**
   Ⓒ **5,280 feet**
   Ⓓ **½ kilometer**

10. **Solve.**
    10 meters = ____

    Ⓐ **1 yard**
    Ⓑ **100 centimeters**
    Ⓒ **10 yards**
    Ⓓ **1 decimeter**

11. **Which of these is the longest?**

    Ⓐ **38 inches**
    Ⓑ **50 centimeters**
    Ⓒ **2 feet**
    Ⓓ **1 yard**

12. Which of these is the heaviest?

Ⓐ ½ pound
Ⓑ 20 ounces
Ⓒ 1 gram
Ⓓ 1/1,000 kilogram

13. Which of these is the greatest volume?

Ⓐ 2 cups
Ⓑ 1 quart
Ⓒ 10 milliliters
Ⓓ ½ gallon

14. Which of these is the shortest?

Ⓐ 6 decimeters
Ⓑ 80 centimeters
Ⓒ ½ meter
Ⓓ 1/10 kilometer

15. Mitchell's tractor has 6 gallons and 4 pints of gasoline in it. He adds 2 gallons and 5 pints more. How much gasoline is now in the tractor?

Ⓐ 9 gallons and 3 pints
Ⓑ 9 gallons and 1 pint
Ⓒ 8 ½ gallons
Ⓓ 9 ½ gallons

16. Jacqueline has a board that is 3 yards long. She saws 2 feet of the board. How long is the board now?

Ⓐ 7 feet
Ⓑ 2 yards and 2 feet
Ⓒ 1 yard
Ⓓ 10 feet

17. Corey has 2 pounds of chocolate chips. He wants to top each cookie with ½ ounce of chocolate chips. How many cookies can he top with chocolate chips?

Ⓐ 64
Ⓑ 16
Ⓒ 40
Ⓓ 50

18. Each section of toy train tracks is 13 centimeters long. If Jude lays 15 sections end-to-end, how long is his train track?

Ⓐ 0.195 kilometers
Ⓑ 195 decimeters
Ⓒ 19.5 centimeters
Ⓓ 1.95 meters

19. Evan has 6 liters of water in a container. He pours out 2,300 milliliters. How much water is left in the container?

Ⓐ 2.3 liters
Ⓑ 3.7 liters
Ⓒ 370 milliliters
Ⓓ 37 liters

20. Ashley has a stick that is 5 ft. 1 in. long. She breaks it into two pieces. If one piece is 1 ft. 10 in. long, how long is the other piece?

Ⓐ 3 ft. 13 in.
Ⓑ 4 ft. 9 in.
Ⓒ 3 ft. 3 in.
Ⓓ 4 ft. 11 in.

21. A bridge has a weight limit of 5,890 pounds. Ralph's truck weighs 3.5 tons. Can he drive his truck over the bridge?

Ⓐ No, because it weighs 35,000 pounds.
Ⓑ No, because it weighs 7,000 pounds.
Ⓒ Yes, because it weighs 3,500 pounds.
Ⓓ Yes, because it weighs 1,750 pounds.

22. It takes Chuck 28 minutes and 12 seconds to complete a swim workout. If he completes 10 swim workouts, how much time has he spent swimming?

Ⓐ 4 hours 42 minutes
Ⓑ 400 minutes
Ⓒ 2 hours 12 minutes
Ⓓ 240 minutes 42 seconds

23. Keith has 7 yards of string. How many inches of string does he have?

Ⓐ 112 inches
Ⓑ 224 inches
Ⓒ 84 inches
Ⓓ 252 inches

24. Which of these is the most reasonable estimate for the total area of the floor space in a house?

Ⓐ 1,200 sq km
Ⓑ 1,200 sq in
Ⓒ 120 sq in
Ⓓ 1,200 sq ft

25. Complete the following.
2.25 hours = _____ minutes

Ⓐ 135
Ⓑ 225
Ⓒ 145
Ⓓ 150

26. The normal body temperature of a person in degrees Celsius is about ______.

Ⓐ 0 degrees Celsius
Ⓑ 37 degrees Celsius
Ⓒ 95 degrees Celsius
Ⓓ 12 degrees Celsius

27. There are 8 pints in a gallon. How many times greater is the volume of a gallon compared to a pint?

Ⓐ 8 times greater
Ⓑ 1/8 times greater
Ⓒ twice as great
Ⓓ 8/10 as great

28. Which of the following measures about 1 dm in length?

Ⓐ a small car
Ⓑ a new crayon
Ⓒ a ladybug
Ⓓ a football field

29. Complete the following.
The area of a postage stamp is about _________ .

Ⓐ 100 sq in
Ⓑ 4 sq in
Ⓒ 10 sq in
Ⓓ 1 sq in

30. Complete the following.
A fully loaded moving truck might weigh ____________ .

Ⓐ **5 tons**
Ⓑ **50 tons**
Ⓒ **500 ounces**
Ⓓ **5,000 ounces**

Representing and Interpreting Data (5.MD.B.2)

1. **A 5th grade science class is raising mealworms. The students measured the mealworms and recorded the lengths on this line plot.**

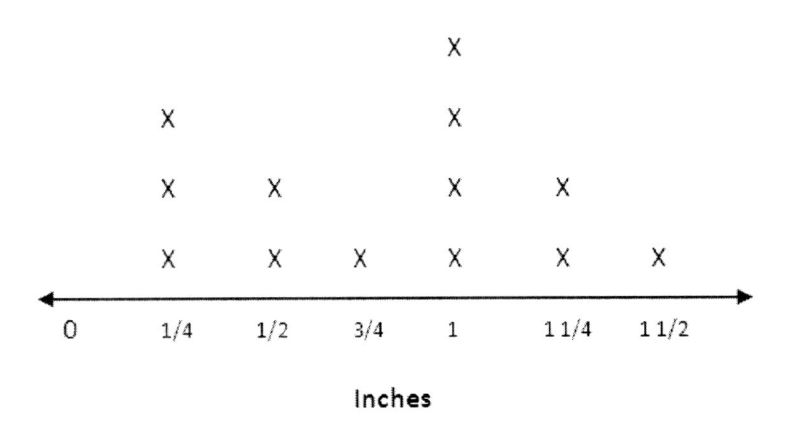

According to this line plot, what was the length of the longest mealworm?

Ⓐ **1/4 inch**
Ⓑ **3/4 inch**
Ⓒ **1 inch**
Ⓓ **1 1/2 inches**

2. **A 5th grade science class is raising mealworms. The students measured the mealworms and recorded the lengths on this line plot.**

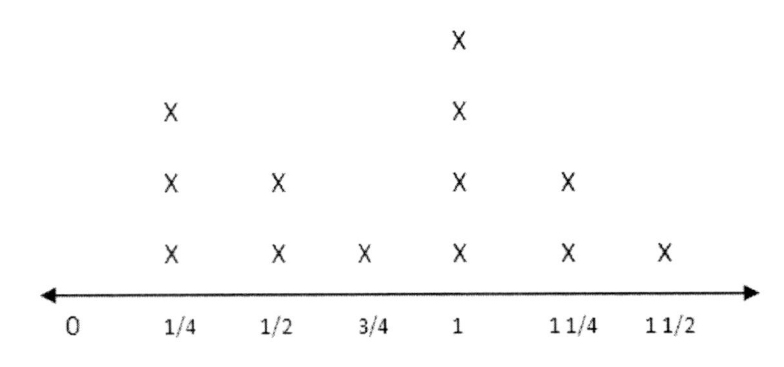

According to this line plot, what was the length of the shortest mealworm?

Ⓐ 1/4 inch
Ⓑ 3/4 inch
Ⓒ 1 1/4 inch
Ⓓ 0

3. A 5th grade science class is raising mealworms. The students measured the mealworms and recorded the lengths on this line plot.

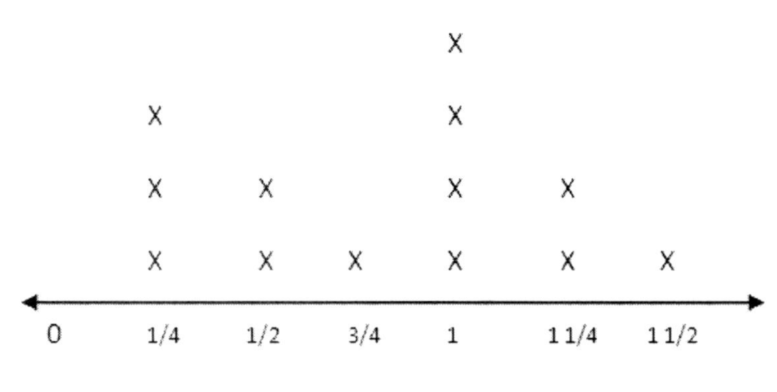

According to this line plot, what was the most common length for mealworms?

Ⓐ 1 1/2 inches
Ⓑ 3/4 inch
Ⓒ 1/4 inch
Ⓓ 1 inch

4. A 5th grade science class is raising mealworms. The students measured the mealworms and recorded the lengths on this line plot.

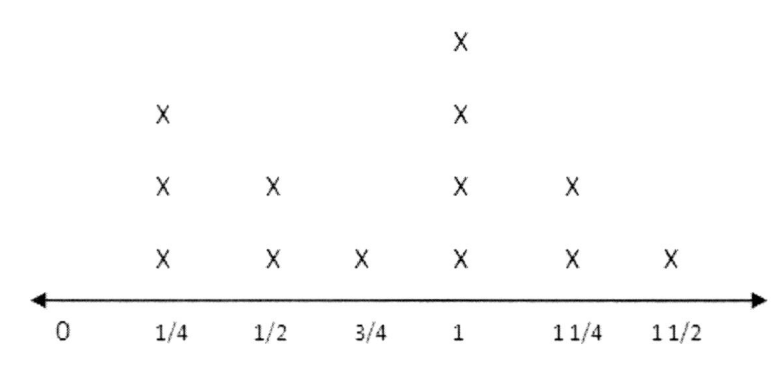

According to this line plot, how many mealworms were less than 1 inch long?

Ⓐ 4
Ⓑ 6
Ⓒ 3
Ⓓ 2

5. A 5th grade science class is raising mealworms. The students measured the mealworms and recorded the lengths on this line plot.

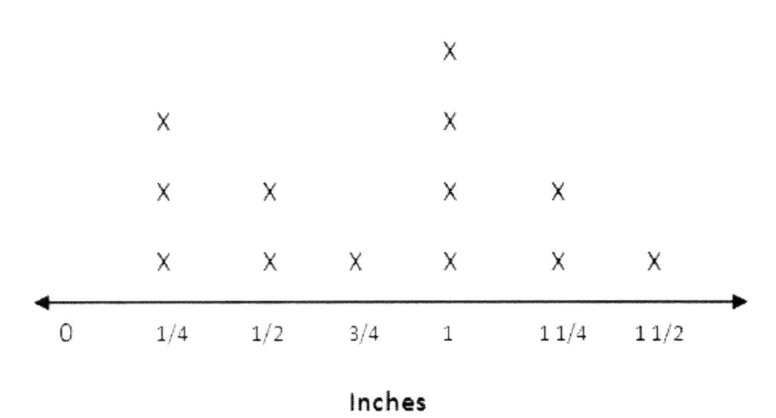

According to this line plot, how many mealworms were measured in all?

Ⓐ 4
Ⓑ 10
Ⓒ 13
Ⓓ 27 3/4

6. A 5th grade science class is raising mealworms. The students measured the mealworms and recorded the lengths on this line plot.

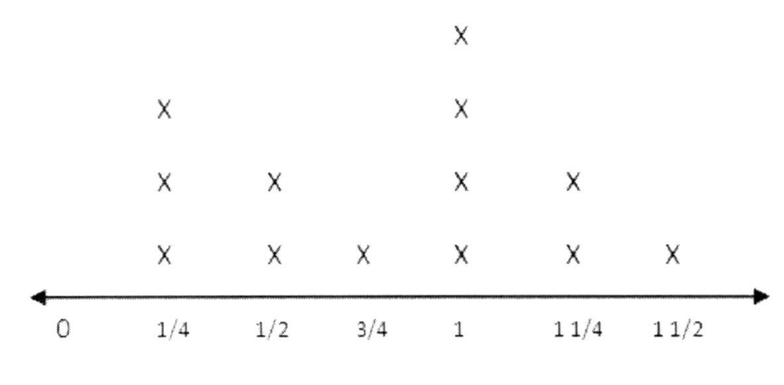

Name: ______________________ Date: ______________

According to this line plot, what is the median length of a mealworm?

Ⓐ 1 inch
Ⓑ 13 inches
Ⓒ between 3/4 inch and 1 inch
Ⓓ 1 1/2 inches

7. A 5th grade science class is raising mealworms. The students measured the mealworms and recorded the lengths on this line plot.

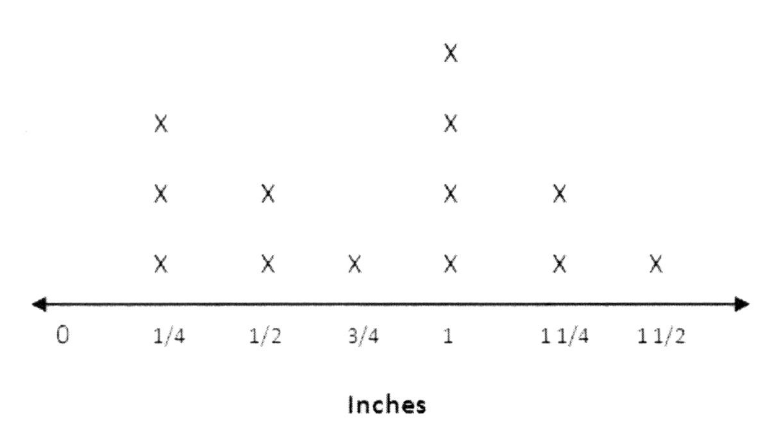

How could someone use this line plot to find the total length of all the mealworms?

Ⓐ Add each of the numbers along the bottom of the line plot
Ⓑ Multiply each of the numbers along the bottom of the line plot
Ⓒ Multiply each length by its number of Xs, then add the values
Ⓓ Multiply each of the numbers along the bottom of the line plot by the total number of Xs, then add the values

8. A 5th grade science class is raising mealworms. The students measured the mealworms and recorded the lengths on this line plot.

Length of Mealworms

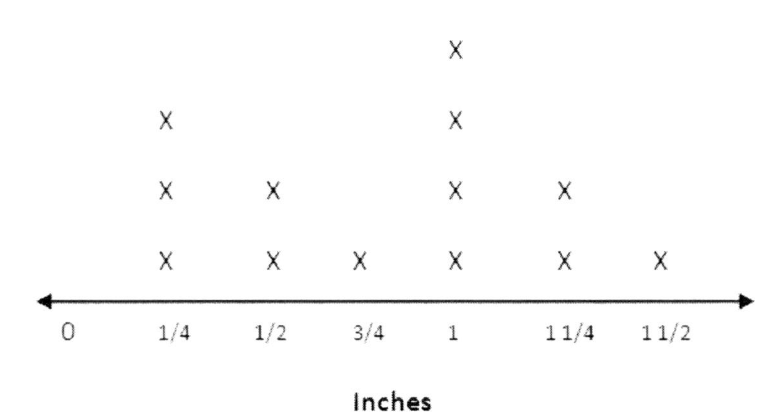

Which of these mealworm lengths would not fit within the range of this line plot?

Ⓐ **7/8 inch**
Ⓑ **1 3/4 inches**
Ⓒ **1 1/8 inches**
Ⓓ **5/16 inch**

9. A 5th grade science class is raising mealworms. The students measured the mealworms and recorded the lengths on this line plot.

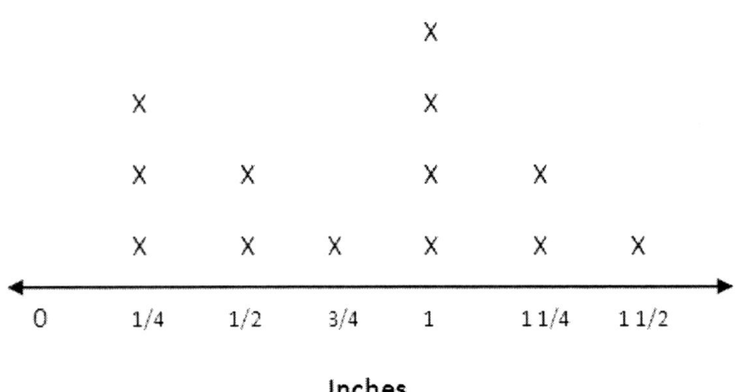

What is the best estimate of the average length of a mealworm according to this plot?

Ⓐ **Just less than an inch**
Ⓑ **About 3 inches**
Ⓒ **Exactly 1 inch**
Ⓓ **Just over an inch**

10. A 5th grade science class is observing weather conditions. The students measured the amount of precipitation each day and recorded it on this line plot.

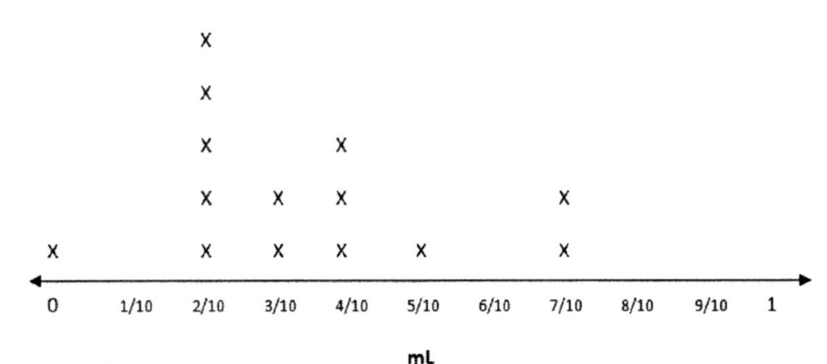

According to this line plot, what was the most precipitation recorded in one day?

Ⓐ **4 7/10 mL**
Ⓑ **7/10 mL**
Ⓒ **9/10 mL**
Ⓓ **2/10 mL**

11. A 5th grade science class is observing weather conditions. The students measured the amount of precipitation each day and recorded it on this line plot.

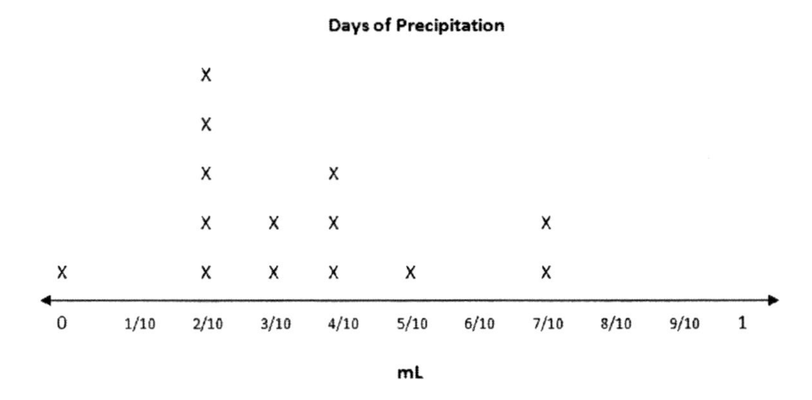

According to this line plot, what was the least amount of precipitation that fell on days that had precipitation?

Ⓐ **1/10 mL**
Ⓑ **6/10 mL**
Ⓒ **0 mL**
Ⓓ **2/10 mL**

12. A 5th grade science class is observing weather conditions. The students measured the amount of precipitation each day and recorded it on this line plot.

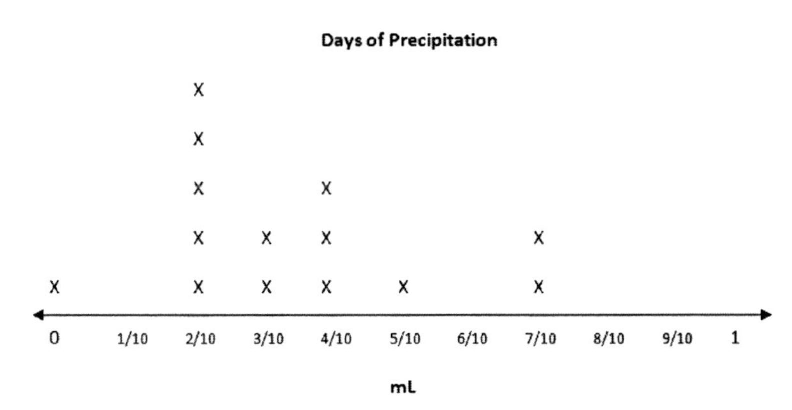

According to this line plot, what was the most common amount of precipitation?

Ⓐ **7/10 mL**
Ⓑ **9/10 mL**
Ⓒ **2/10 mL**
Ⓓ **0 mL**

13. A 5th grade science class is observing weather conditions. The students measured the amount of precipitation each day and recorded it on this line plot.

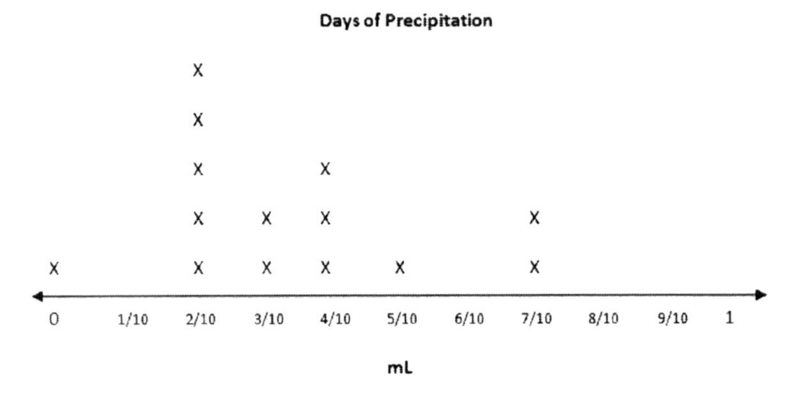

According to this line plot, how many days received less than 3/10 mL of precipitation?

Ⓐ **6**
Ⓑ **0**
Ⓒ **4**
Ⓓ **3**

14. A 5th grade science class is observing weather conditions. The students measured the amount of precipitation each day and recorded it on this line plot.

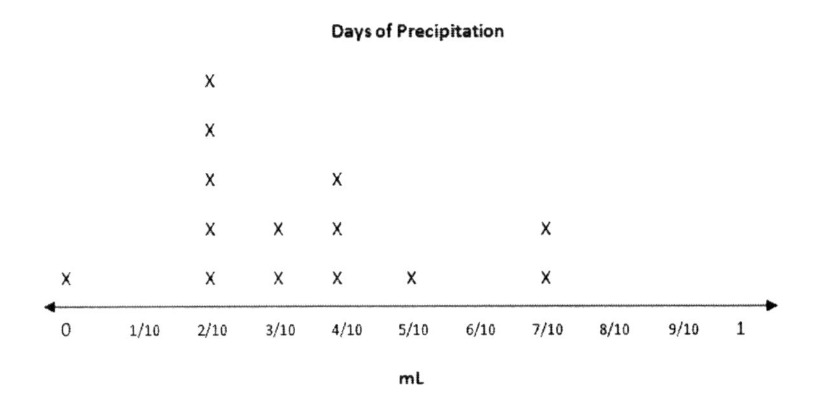

According to this line plot, how many days were observed in all?

Ⓐ 13
Ⓑ 5
Ⓒ 14
Ⓓ 11

15. A 5th grade science class is observing weather conditions. The students measured the amount of precipitation each day and recorded it on this line plot.

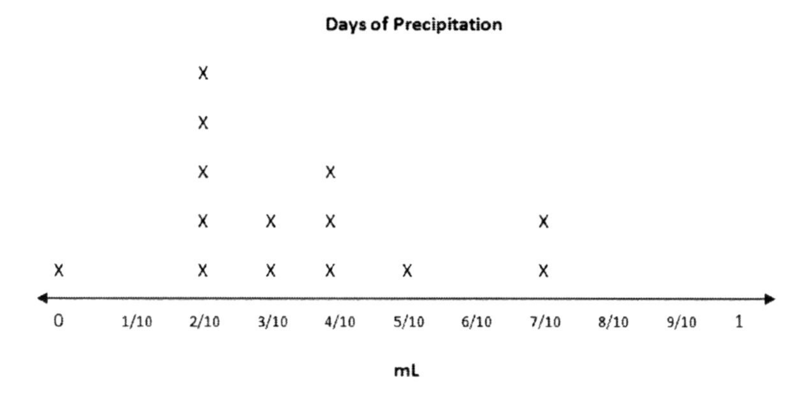

According to this line plot, what is the median amount of precipitation?

Ⓐ 7/10 mL
Ⓑ 3/10 mL
Ⓒ 2/10 mL
Ⓓ 0 mL

16. A 5th grade science class is observing weather conditions. The students measured the amount of precipitation each day and recorded it on this line plot.

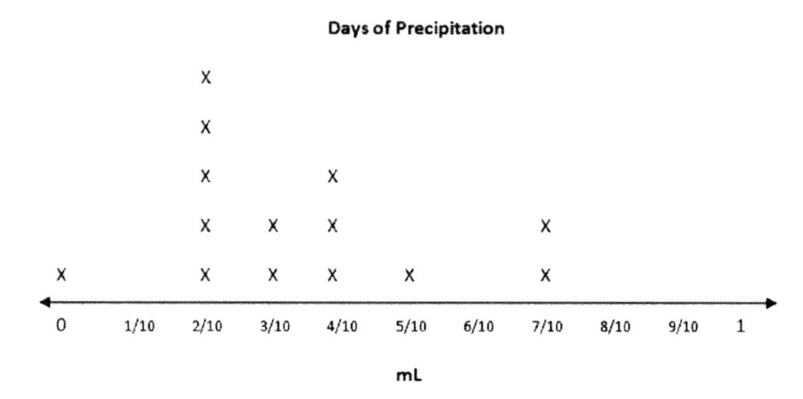

How could someone use this line plot to find the total amount of all the precipitation that fell?

Ⓐ Add each of the numbers along the bottom of the line plot
Ⓑ Multiply each of the numbers along the bottom of the line plot by the total number of Xs, then add the values
Ⓒ Multiply each of the numbers along the bottom of the line plot
Ⓓ Multiply each measurement by its number of Xs, then add the values

17. A 5th grade science class is observing weather conditions. The students measured the amount of precipitation each day and recorded it on this line plot.

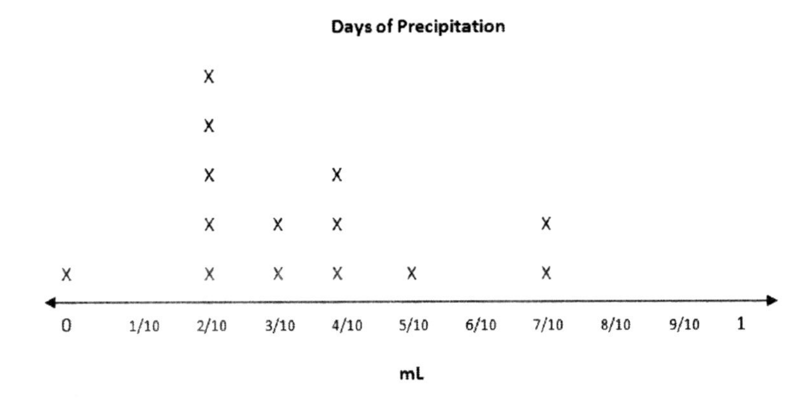

Which of these measurements would not fit within the range of this line plot?

Ⓐ 12/10 mL
Ⓑ 9/10 mL
Ⓒ 0 mL
Ⓓ 2/10 mL

18. A 5th grade science class is observing weather conditions. The students measured the amount of precipitation each day and recorded it on this line plot.

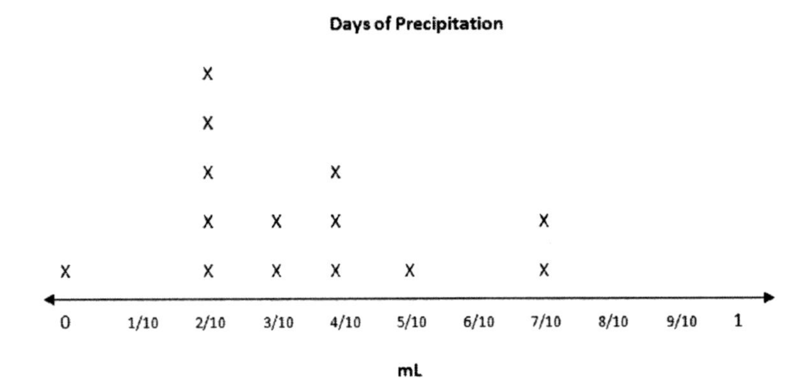

Suppose the class recorded data for one more day, but there was no precipitation. What should they do?

Ⓐ **Leave the line plot as it is**
Ⓑ **Add an X above the one in the 0 column**
Ⓒ **Erase one of the Xs from the line plot**
Ⓓ **Make up a value and put an X in that column**

19. A 5th grade science class is observing weather conditions. The students measured the amount of precipitation each day and recorded it on this line plot.

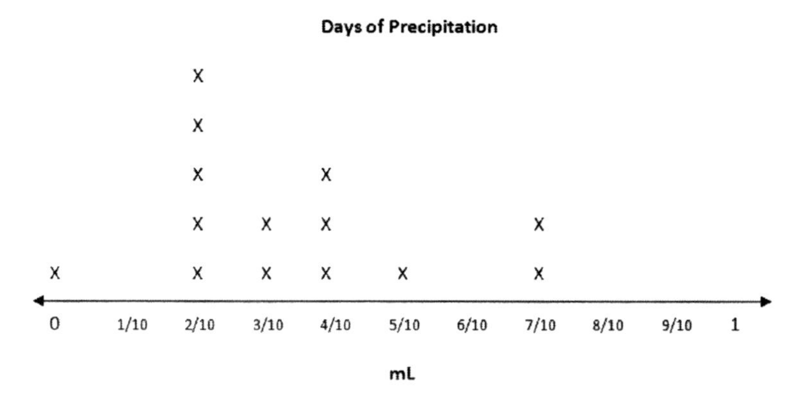

How could someone find the average amount of precipitation that fell during this time period?

Ⓐ **Subtract the lowest value from the highest value, and then multiply by the total number of Xs.**
Ⓑ **Move the Xs so there are an equal number above each value, then multiply that number by the median.**
Ⓒ **Multiply each value by the number of Xs above it, add those values together, and then divide by the total number of Xs.**
Ⓓ **Count the Xs from the lowest and highest values toward the middle of the plot and determine the value of the X that falls in the very middle.**

20. A 5th grade science class went on a nature walk. The students each selected one leaf and weighed it when they got back to the room. They recorded their data on this line plot.

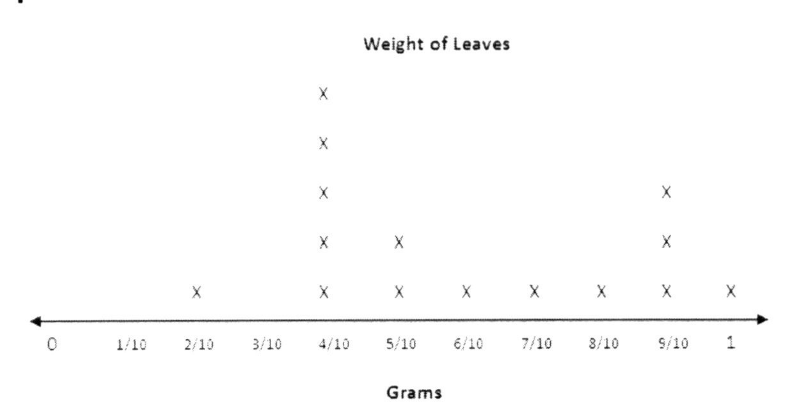

According to this line plot, which is the least frequent weight of a leaf?

Ⓐ 7/10 g
Ⓑ 9/10 g
Ⓒ 4/10 g
Ⓓ 5/10 g

21. A 5th grade science class went on a nature walk. The students each selected one leaf and weighed it when they got back to the room. They recorded their data on this line plot.

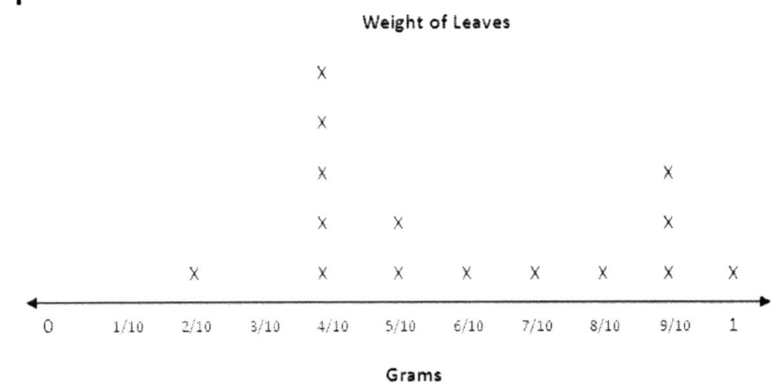

According to this line plot, what is the range of weights for these leaves?

Ⓐ 8/10 g
Ⓑ 1 g
Ⓒ 5/10 g
Ⓓ 4/10 g

22. A 5th grade science class went on a nature walk. The students each selected one leaf and weighed it when they got back to the room. They recorded their data on this line plot.

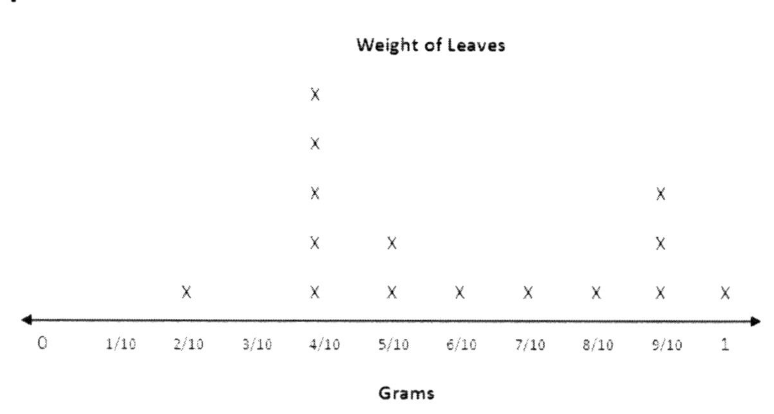

According to this line plot, what is the mode for this set of data?

Ⓐ 1
Ⓑ 15
Ⓒ 5/10
Ⓓ 4/10

23. A 5th grade science class went on a nature walk. The students each selected one leaf and weighed it when they got back to the room. They recorded their data on this line plot.

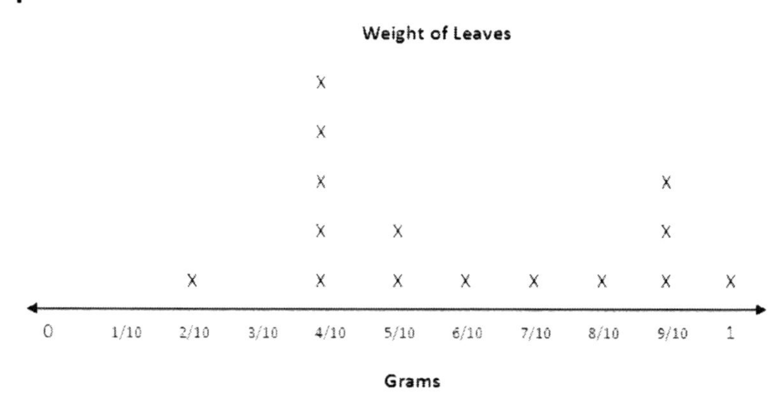

According to this line plot, how many leaves weigh more than 7/10 g?

Ⓐ 1
Ⓑ 5
Ⓒ 9
Ⓓ 0

24. A 5th grade science class went on a nature walk. The students each selected one leaf and weighed it when they got back to the room. They recorded their data on this line plot.

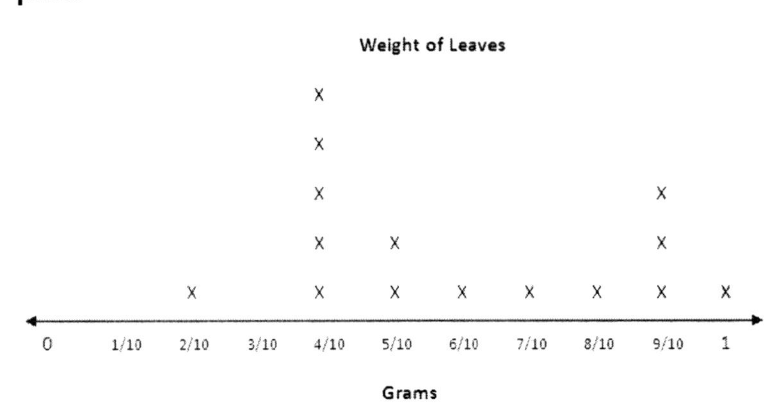

According to this line plot, how many leaves were measured in all?

Ⓐ 15
Ⓑ 4/10
Ⓒ 6/10
Ⓓ 11

25. A 5th grade science class went on a nature walk. The students each selected one leaf and weighed it when they got back to the room. They recorded their data on this line plot.

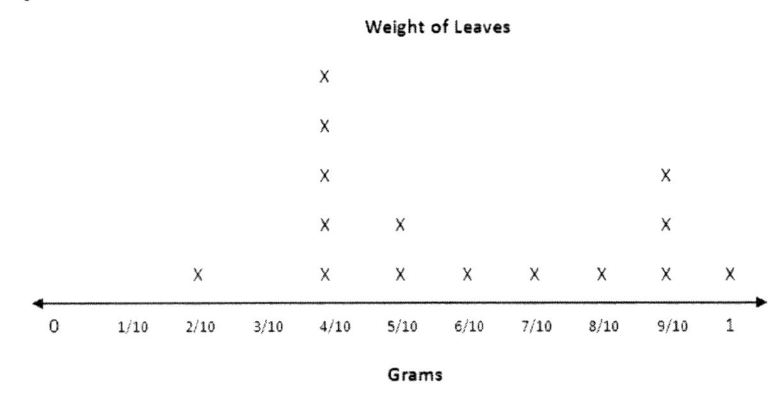

According to this line plot, how many leaves weigh less than the most frequent weight?

Ⓐ 9
Ⓑ 5
Ⓒ 0
Ⓓ 1

26. **A 5th grade science class went on a nature walk. The students each selected one leaf and weighed it when they got back to the room. They recorded their data on this line plot.**

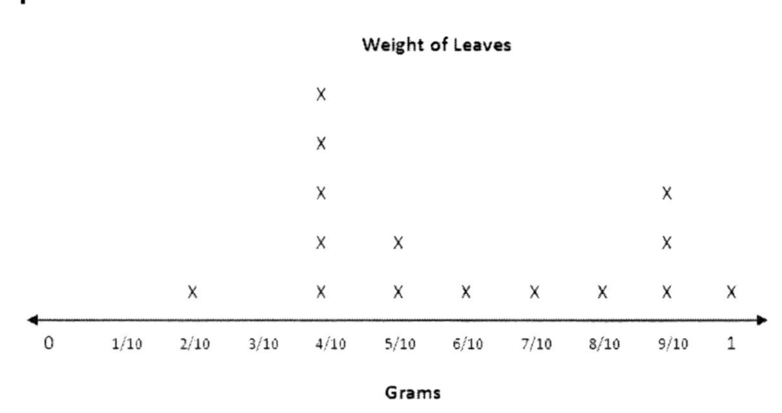

What is the total weight of all the leaves?

Ⓐ **9 g**
Ⓑ **9/10 g**
Ⓒ **90 g**
Ⓓ **9/100 g**

27. **A 5th grade science class went on a nature walk. The students each selected one leaf and weighed it when they got back to the room. They recorded their data on this line plot.**

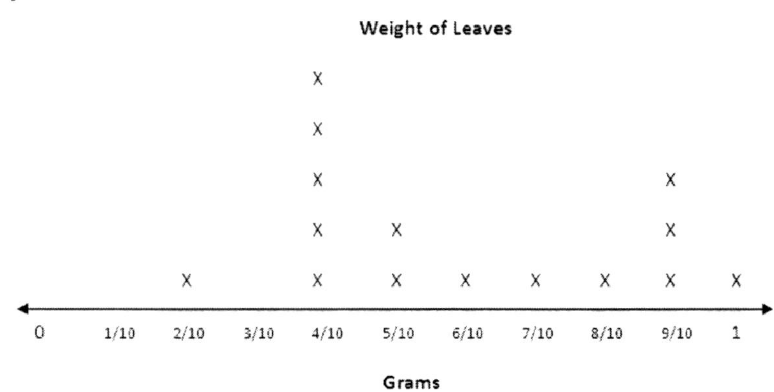

What is the average weight of the leaves?

Ⓐ **5/10 g**
Ⓑ **9/10 g**
Ⓒ **6/10 g**
Ⓓ **8/10 g**

28. A 5th grade science class went on a nature walk. The students each selected one leaf and weighed it when they got back to the room. They recorded their data on this line plot.

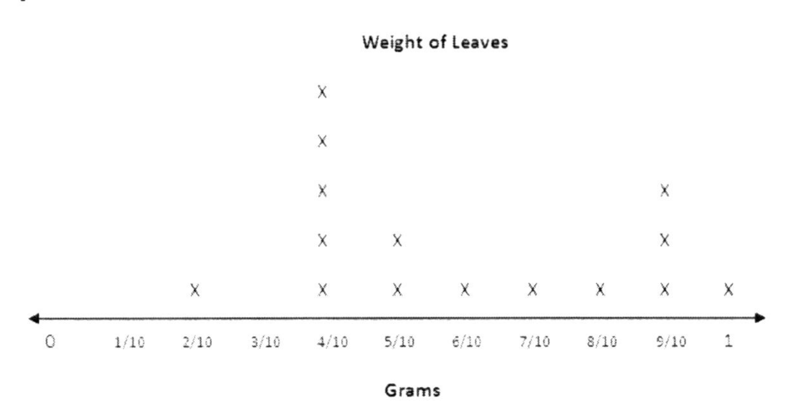

Which of these is greatest for this set of data?

Ⓐ The range
Ⓑ The mean
Ⓒ The median
Ⓓ The mode

29. Which of these best describes the mode of a set of data?

Ⓐ The value that falls in the middle of all the data points
Ⓑ The value that has the most data points
Ⓒ The difference between the highest and lowest data points
Ⓓ The average of all the data points

30. Which of these best describes the mean of a set of data?

Ⓐ The difference between the highest and lowest data points
Ⓑ The average of all the data points
Ⓒ The value that has the most data points
Ⓓ The value that falls in the middle of all the data points

Name: ______________________ Date: ______________

Geometric Measurement: Understand Concepts of Volume

Volume (5.MD.C.3.A)

1. Which type of unit might be used to record the volume of a rectangular prism?

Ⓐ inches
Ⓑ square inches
Ⓒ ounces
Ⓓ cubic inches

2. Maeve needed to pack a crate that measured 4 ft. by 2 ft. by 3 ft. with 1 foot cubes. How many 1 foot cubes can she fit in the crate?

Ⓐ 12
Ⓑ 48
Ⓒ 24
Ⓓ 9

3. The volume of an object is the amount of ______________________.

Ⓐ space it occupies
Ⓑ dimensions it has
Ⓒ layers you can put in it
Ⓓ weight it can hold

4. Which of these could be filled with about 160 cubes of sugar if each sugar cube is one cubic centimeter?

Ⓐ a ring box
Ⓑ a moving box
Ⓒ a cereal box
Ⓓ a sandbox

5. Tony and Yolani are measuring the volume of a supply box at school. Tony uses a ruler to measure the box's length, width, and height in centimeters; then he mul tiplies these measurements. Yolani fills the box with centimeter cubes, then counts the number of cubes. How will their answers compare?

Ⓐ They cannot be compared because they used different units.
Ⓑ They will be almost or exactly the same.
Ⓒ Tony's answer will be greater than Yolani's.
Ⓓ Yolani's answer will be greater than Tony's.

Cubic Units (5.MD.C.3.B)

1. Stan covers the bottom of a box with 8 centimeter cubes, leaving no gaps. He is able to build 4 layers of cubes to fill the box completely. What is the volume of the box?

Ⓐ 32 centimeters
Ⓑ The square of 32 cm
Ⓒ 32 cm^2
Ⓓ 32 cubic centimeters

2. Which of these is an accurate way to measure the volume of a rectangular prism?

Ⓐ Fill it with water and then weigh the water
Ⓑ Trace each face of the prism on centimeter grid paper, and then count the number of squares it comprises
Ⓒ Measure the length and the width, and then multiply the two values
Ⓓ Pack it with unit cubes, leaving no gaps or overlaps, and count the number of unit cubes

3. Which of these could possibly be the volume of a cereal box?

Ⓐ 360 in^3
Ⓑ 520 sq cm
Ⓒ 400 cubic feet
Ⓓ 385 dm^2

4. A container measures 4 inches wide, 6 inches long, and 10 inches high. How many 1 inch cubes will it hold?

Ⓐ 20^2
Ⓑ 240
Ⓒ The cube of 240
Ⓓ Cannot be determined

5. Annie covers the bottom of a box with 6 centimeter cubes, leaving no gaps. If the volume of the box is 30 cm^3, how many more centimeter cubes will she be able to fit inside?

Ⓐ 5
Ⓑ 180
Ⓒ 24
Ⓓ 4

Counting Cubic Units (5.MD.C.4)

1. **What is the volume of the figure?**

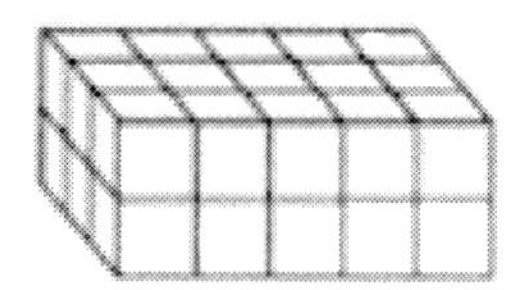

Ⓐ **60 cubic units**
Ⓑ **15 cubic units**
Ⓒ **30 cubic units**
Ⓓ **31 cubic units**

2. **What is the volume of the figure?**

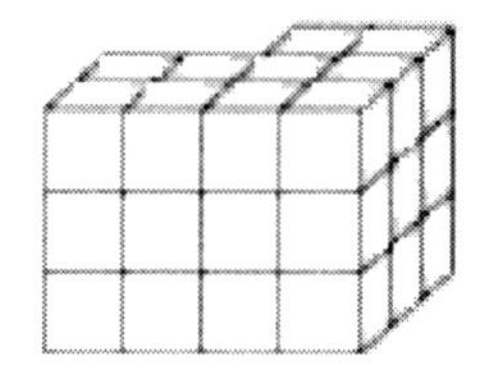

Ⓐ **30 units³**
Ⓑ **27 units³**
Ⓒ **31 units³**
Ⓓ **36 units³**

3. **Which of these has a volume of 24 cubic units?**

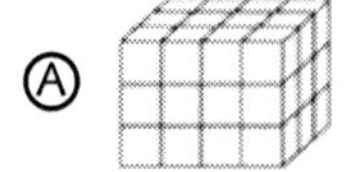

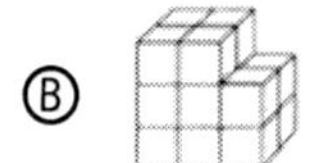

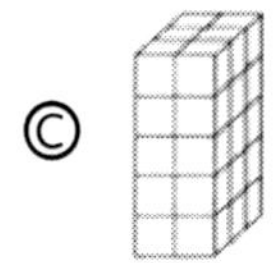

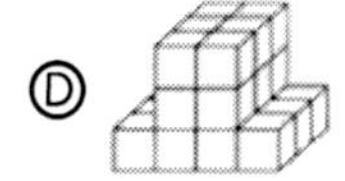

4. **Trevor is building a tower out of centimeter cubes. This is the base of the tower so far.**

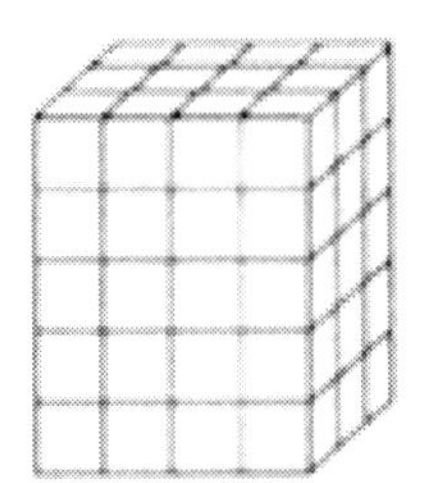

How many more layers must Trevor add to have a tower with a volume of 84 cm^3?

Ⓐ 7
Ⓑ 2
Ⓒ 5
Ⓓ 4

5. **Kerry built the figure on the left and Milo built the one on the right. If they knock down their two figures to build one large one using all of the blocks, what will its volume be?**

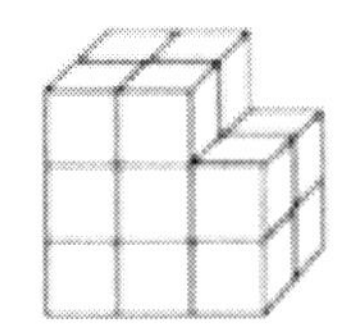

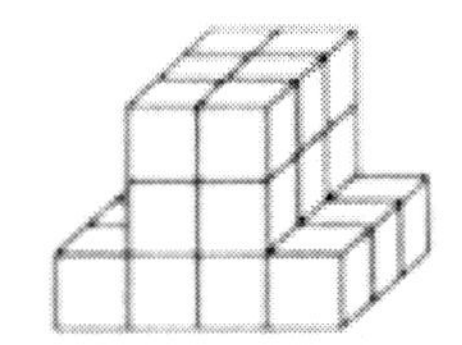

Ⓐ 34 cubic units
Ⓑ 16 cubic units
Ⓒ 52 cubic units
Ⓓ 40 cubic units

Multiply to Find Volume (5.MD.C.5.A)

1. **What is the volume of the figure?**

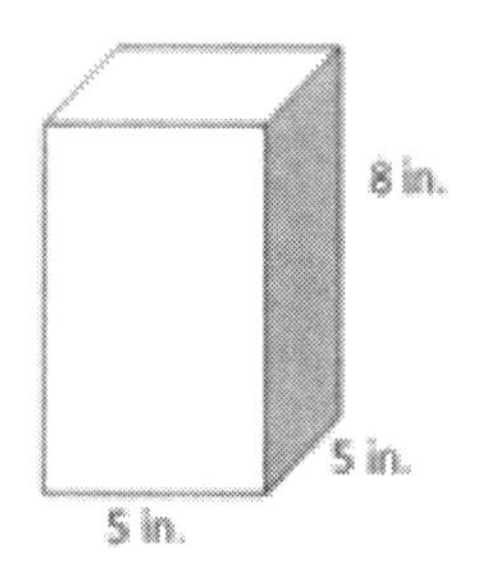

Ⓐ **33 in^3**
Ⓑ **18 in^3**
Ⓒ **80 in^3**
Ⓓ **200 in^3**

2. **What is the volume of the figure?**

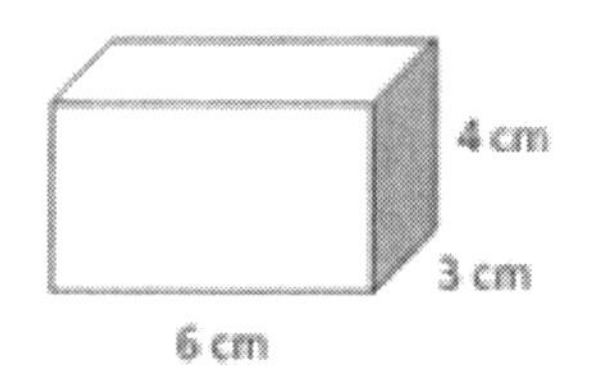

Ⓐ **13 cm^3**
Ⓑ **22 cm^3**
Ⓒ **72 cm^3**
Ⓓ **700 cm^3**

3. **The figure has a volume of 66 ft^3. What is the height of the figure?**

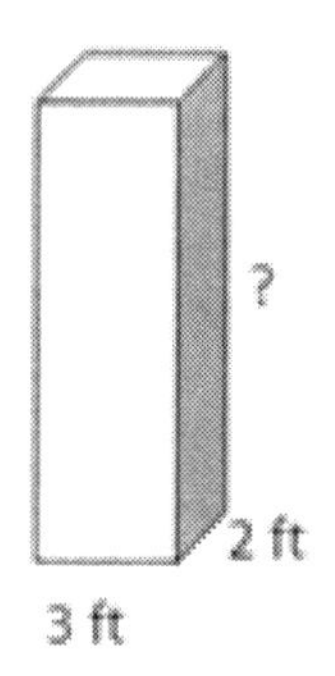

Ⓐ **11 ft**
Ⓑ **61 ft**
Ⓒ **13 ft**
Ⓓ **33 ft**

4. The figure has a volume of 14 in³. What is the width of the figure?

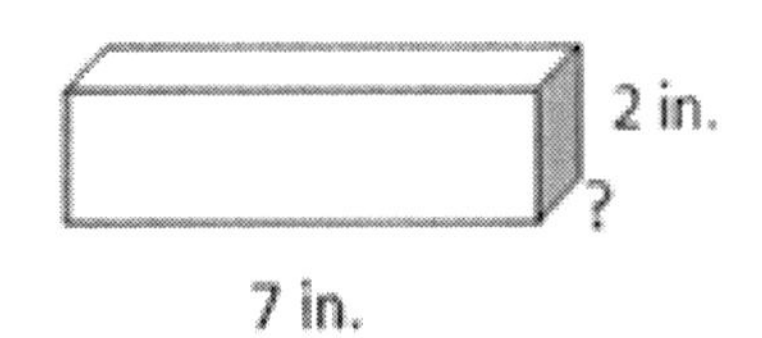

Ⓐ **2 in.**
Ⓑ **1 in.**
Ⓒ **5 in.**
Ⓓ **2.5 in.**

5. Which figure has an area of 42 m³?

Ⓐ 3 m, 2 m, 6 m

Ⓑ 5 m, 5 m, 5 m

Ⓒ 9 m, 2 m, 4 m

Ⓓ 2 m, 3 m, 7 m

Real World Problems with Volume (5.MD.C.5.B)

1. Michael packed a box full of 1 ft cubes. The box held 54 cubes. Which of these could be the box Michael packed?

Ⓐ 

Ⓑ 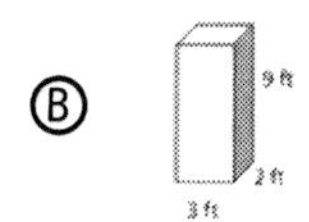

Ⓒ 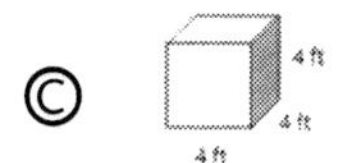

Ⓓ 

2. A container is shaped like a rectangular prism. The area of its base is 30 in^2. If the container is 5 inches tall, how many 1 inch cubes can it hold?

Ⓐ 150
Ⓑ 35
Ⓒ 4500
Ⓓ 95

3. A rectangular prism has a volume of 300 cm^3. If the area of its base is 25 cm^2 how tall is the prism?

Ⓐ 325 cm
Ⓑ 7500 cm
Ⓒ 12 cm
Ⓓ 275 cm

4. Antonia wants to buy a jewelry box with the greatest volume. She measures the length, width, and height of four different jewelry boxes. Which one should she buy to have the greatest volume?

Ⓐ 10 in x 7 in x 4 in
Ⓑ 8 in x 5 in x 5 in
Ⓒ 12 in x 5 in x 5 in
Ⓓ 14 in x 2 in x 10 in

5. Damien is building a file cabinet that must hold 20 ft^3. He has created a base for the cabinet that is 4 ft by 1 ft. How tall should he build the cabinet?

Ⓐ 25 ft
Ⓑ 20 ft
Ⓒ 4 ft
Ⓓ 5 ft

Adding Volumes (5.MD.C.5.C)

1. A refrigerator has a 3 foot by 2 foot base. The refrigerator portion is 4 feet high and the freezer is 2 feet high. What is the total volume?

Ⓐ 36 ft^3
Ⓑ 26 ft^3
Ⓒ 11 ft^3
Ⓓ 16 ft^3

2. Matthew has two identical coolers. Each one measures 30 inches long, 10 inches wide, and 15 inches high. What is the total volume of the two coolers?

Ⓐ 4,500 in^3
Ⓑ 9,000 in^3
Ⓒ 110 in^3
Ⓓ 3,025 in^3

3. Ingrid is packing 1 foot square boxes into shipping crates. She has two shipping crates, shown below. How many boxes can she pack in them all together?

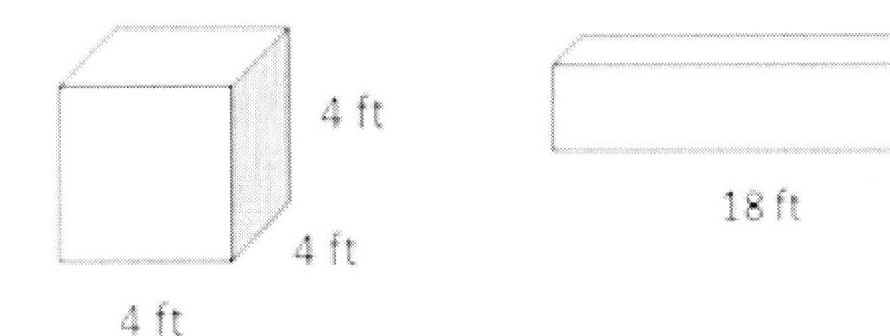

Ⓐ 64
Ⓑ 33
Ⓒ 82
Ⓓ 100

4. Bryson has two identical bookcases stacked one on top of the other. Together, they hold 48 ft^3. If the area of the base is 8 ft^2, how tall is each bookcase?

Ⓐ 40 ft
Ⓑ 6 ft
Ⓒ 3 ft
Ⓓ 20 ft

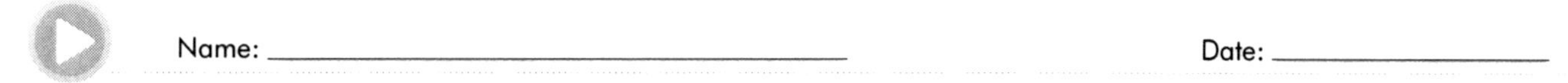

5. **Amy built a house for her gerbil out of two boxes. One box measures 6 cm by 3 cm by 10 cm and the other measures 4 cm by 2 cm by 2 cm. What is the total volume of the gerbil house?**

Ⓐ **196 cm^3**
Ⓑ **180 cm^3**
Ⓒ **27 cm^3**
Ⓓ **2,880 cm^3**

End of Measurement and Data

Measurement and Data

Answer Key
&
Detailed Explanations

Converting Units of Measure (5.MD.A.1)

Question No.	Answer	Detailed Explanation
1	C	There are 12 inches in a foot.
2	B	There are 16 ounces in a pound.
3	D	There are 100 centimeters in a meter.
4	A	There are 4 quarts in a gallon.
5	B	There are 3 feet in a yard.
6	C	There are 1,000 grams in a kilogram.
7	A	There are 60 seconds in a minute and 60 minutes in an hour. Therefore, 60 x 60 = 3,600.
8	D	There are 1,000 milliliters in a liter.
9	C	There are 5,280 feet in a mile.
10	D	There are 10 meters in a decimeter.
11	A	There are 36 inches or 3 feet in a yard, so 38 inches is longer than 1 yard or 2 feet. It is also longer than 50 centimeters, which is half a meter (close to half a yard).
12	B	There are 16 ounces in a pound, so 20 ounces is more than ½ pound. A single ounce is much heavier than 1 gram (or 1/1,000 kilogram), so 20 ounces is heavier still.
13	D	There are 16 cups, or 4 quarts, in a gallon, so ½ gallon is more than 2 cups or 1 quart. It is much more than 10 milliliters, since even a liter (1,000 milliliters) is less than ½ gallon.
14	C	There are 100 centimeters in a meter, therefore 80 centimeters is more than ½ meter. There are 10 decimeters in a meter, therefore 6 decimeters is more than ½ meter. There are 1,000 meters in a kilometer, therefore 1/10 kilometer is 100 meters.
15	B	6 gal 4 pt + 2 gal 5 pt = 8 gal 9 pt. Since there are 8 pints in a gallon, this can be converted to 9 gallons and 1 pint.
16	A	There are 3 feet in a yard, so the board began as 9 feet long. After sawing off two feet, she is left with a board 7 feet long (9 – 7 = 2).

Question No.	Answer	Detailed Explanation
17	A	There are 16 ounces in a pound, so Corey has 32 ounces of chips (16 x 2 = 32). Divide by ½ ounce to solve: 32 ÷ ½ = 32 x 2 = 64
18	D	Multiply to find the length of the track (13 cm x 15 = 195 cm). Since there are 100 centimeters in a meter, divide by 100 (move the decimal two places to the left) to convert the answer to meters.
19	B	There are 1,000 milliliters in a liter, so Evan began with 6,000 liters. 6,000 – 2,300 = 3,700 To convert milliliters back to liters, divide by 1,000 (move the decimal 3 places to the left).
20	C	Since there are 12 inches in a foot, you can think of 5 ft. 1 in. as 4 ft. 13 in. (by converting one of the feet to inches). From that, subtract 1 ft. 10 in. to get 3 ft. 3 in.
21	B	There are 2,000 pounds in a ton. Therefore, Ralph's truck weighs 7,000 pounds (3.5 x 2,000 = 7,000), which is too heavy for the bridge.
22	A	28 min 12 sec multiplied by 10 is 280 min 120 sec. The 120 sec can be converted to 2 min, for a total of 282 min. Since there are 60 minutes in an hour, 240 of those 282 minutes can be converted to 4 hours (60 x 4 = 240), leaving 42 minutes more (282 -240 = 42).
23	D	There are 12 inches in foot and 3 feet in a yard. Therefore, each yard is made up of 36 inches (3 x 12 = 36). Seven yards would be 7 x 36 inches, which is 252.
24	D	Square inches are relatively small units of measure. Even 1,200 of them would be smaller than the area of one room. Square kilometers are relatively large units of measure. Even one sq km would be larger than a house. The only reasonable answer is 1,200 sq ft.

Question No.	Answer	Detailed Explanation
25	A	There are 60 minutes in each hour, so multiply: 2.25 x 60 300 1200 12000 135.00
26	B	At 0 degrees Celsius, water freezes. At 100 degrees Celsius, water boils. The only reasonable temperature for a human body is somewhere in between. The best estimate is 37 degrees Celsius.
27	A	8 pints =1 gallon, a gallon is 8 times greater in volume than a pint.
28	B	The prefix deci means ten. One dm (decimeter) is 10 centimeters in length.
29	D	The area of an object is equal to its length x width. A postage stamp is about 1 in long and 1 in wide, so its area is about 1 sq in. The other answer options would all be much larger than a postage stamp.
30	A	Ounces are unreasonable estimates for this scenario, even if there are 5,000 of them (that's about 300 pounds). Loaded trucks are measured in tons, with one ton equal to 2,000 pounds. Five tons is equal to 10,000 pounds (5 x 2,000), which is a reasonable estimate for a loaded truck. The option 50 tons (50 x 2,000) is equal to 100,000 pounds, which is much too large.

Representing and Interpreting Data (5.MD.B.2)

Question No.	Answer	Detailed Explanation
1	D	The length of each mealworm is shown along the bottom of the line plot. The highest value on the scale is 1 1/2 inches and the Xs above show that there were mealworms this long.
2	A	The length of each mealworm is shown along the bottom of the line plot. The lowest value on the scale is 1/4 inch and the Xs above show that there were mealworms this long.
3	D	The Xs on the line plot represent the number of mealworms at each length. Since 1 inch has the most Xs above it (4), it is the most common length.

Question No.	Answer	Detailed Explanation
4	B	The Xs on the line plot represent the number of mealworms at each length. There were 3 mealworms that were 1/4 inch long, 2 that were 1/2 inch long, and 1 that was 3/4 inch long. Altogether, that's 6 mealworms that are less than 1 inch long.
5	C	The Xs on the line plot represent the number of mealworms at each length. There are 13 Xs in all, at various lengths.
6	A	The median of a set of data is the middle value. In this set of 13 lengths, the middle value would be one of the 1 inch lengths.
7	C	To find the total length, you would have to add together the three mealworms that are 1/4 inch long (3 x 1/4) plus the two that are 1/2 inch long (2 x 1/2) and so on for each length.
8	B	1 3/4 inches is greater than 1 1/2 inches, so it would not fall within the range of 1/4 to 1 1/2 inches as all of the other measurements.
9	A	To find the average length, first find the length of all the mealworms combined: $(3 \times \frac{1}{4}) + (2 \times \frac{1}{2}) + (1 \times \frac{3}{4}) + (4 \times 1) + (2 \times 1\frac{1}{4}) + (1 \times 1\frac{1}{2}) =$ $\frac{3}{4} + 1 + \frac{3}{4} + 4 + 2\frac{1}{2} + 1\frac{1}{2} =$ $\frac{3}{4} + 1 + \frac{3}{4} + 4 + 2\ 2/4 + 1\ 2/4 =$ $8\ 10/4 =$ $10\ 2/4 =$ $10\frac{1}{2}$ Then divide the total length by the number of mealworms measured: $10\frac{1}{2} \div 13 =$ $10\frac{1}{2} \times 1/13 =$ $21/2 \times 1/13 =$ $21/26$
10	B	The amount of precipitation each day is shown along the bottom of the line plot. The highest value on the scale that has Xs above it, showing that there was a day that received that amount, is 7/10 mL.
11	D	The amount of precipitation each day is shown along the bottom of the line plot. The lowest value on the scale other than zero that has Xs above it is 2/10 mL.

Question No.	Answer	Detailed Explanation
12	C	The Xs on the line plot represent the number of days that received that much precipitation. Since 2/10 mL has the most Xs above it (5 Xs), it is the most common amount.
13	A	The Xs on the line plot represent the number of days that received that amount of precipitation. There was 1 day that received 0 mL and 5 days that received 2/10 mL. Altogether, that's 6 days that received less than 3/10 mL of precipitation.
14	C	The Xs on the line plot represent the number of days that received a certain amount of precipitation. There are 14 Xs in all, at various amounts.
15	B	The median of a set of data is the middle value. In this set of 14 data points, the middle value would be one of the 3/10 mL recordings.
16	D	To find the total amount, you would have to add together the 5 days that received 2/10 mL (5 x 2/10) plus the 2 that received 3/10 mL (2 x 3/10) and so on for each data point.
17	A	12/10 mL is greater than 1 mL, so it would not fall within the range of 0 to 1 mL as all of the other measurements do.
18	B	All measurements need to be recorded. If there was 0 precipitation on that day, students should mark an X above the one in the 0 column.
19	C	The average of a set of data is found by adding all of the values together and dividing by the number of data points.
20	A	The value 7/10 g has only one X above it, which means only one leaf weighed this much. The other values given have more than one X above them, indicating that more than one leaf had that weight. This makes 7/10 the least frequent of these options.
21	A	The range for a set of data is the difference between the highest and lowest data points. On this line plot, the highest value is 1 and the lowest value is 2/10, so the range is 8/10.
22	D	The mode for a set of data is the value that occurs most frequently. Since 4/10 g has the most Xs above it 5 total, it is the most frequent weight.

Question No.	Answer	Detailed Explanation
23	B	The Xs on the line plot represent the number of leaves at each weight. There is one leaf that weighs 8/10 g, three that weigh 9/10 g, and one that weighs 1 g. Altogether, that's 5 leaves that weigh more than 7/10 g.
24	A	The Xs on the line plot represent the number of leaves at each weight. There are 15 Xs in all, at various weights.
25	D	The most frequent weight for this set of data is 4/10 g, which has 5 data points. The only leaf that weighs less than these is the one leaf that weighs 2/10 g.
26	A	To find the total weight, you would have to add together the weight of the one leaf that is 2/10 g (1 x 2/10) plus the five that are 4/10 g (5 x 4/10) and so on for each data point. The total will be 90/10 g, which is equal to 9 g.
27	C	To find the average weight of the leaves, add the weight of all 15 leaves to get 9 grams. Divide this amount by the number of leaves (15) to get 6/10 g.
28	A	The range (highest value minus lowest value) is 8/10, which is greater than the mean (average = 6/10), median (middle = 5/10), and mode (most frequent 4/10).
29	B	The mode is the most frequently occurring value. Therefore, it is the one with the most data points (or most Xs) above it.
30	B	The mean of a set of data is just another word for the average.

Volume (5.MD.C.3.A)

Question No.	Answer	Detailed Explanation
1	D	To find the volume of a rectangular solid, multiply the area of the base (l x w) by the height (h). Therefore, the units are cubic units of length, such as cubic inches.
2	C	To find the volume of a rectangular solid, multiply the area of the base (l x w) by the height (h). In this problem, 4 x 2 x 3 = 24 cubic feet. Therefore, it will take 24 cubes to fill the crate, since each cube is one cubic foot.

Question No.	Answer	Detailed Explanation
3	A	Volume is a measurement of the space an object occupies. It is measured in cubic units.
4	C	The formula for determining volume is l x w x h. A cereal box could be about 8 cm x 2 cm x 10 cm, which is 160 cm3.
5	B	The formula for determining volume is l x w x h. It can also be determined by counting the number of unit cubes that fill a solid figure. Since Tony and Yolani both used centimeters as their units, their two methods should give them almost the same answer.

Cubic Units (5.MD.C.3.B)

Question No.	Answer	Detailed Explanation
1	D	The volume of a container is measured in cubic units (or units^3).
2	D	An object's volume can be determined by packing it with unit cubes, leaving no gaps or overlaps, and counting the number of unit cubes.
3	A	This is the best option because not only is the value reasonable (a cereal box could measure 3 in x 10 in x 12 in, for example), but it also uses units appropriate for measuring volume.
4	B	The volume of the container is 240 in^3 (4 x 6 x 10). That means it can hold 240 1-inch cubes.
5	C	If the volume of the box is 30 cm^3, it can hold 30 centimeter cubes. Since there are already 6 in the box, there is room for 24 more (30 – 6 = 24).

Counting Cubic Units (5.MD.C.4)

Question No.	Answer	Detailed Explanation
1	C	The figure clearly has 15 cubes in the top layer, so there must be another 15 cubes in the bottom layer (the figure is only 2 units high, or 2 layers). Therefore, it has a volume of 30 cubic units (15 + 15 = 30).
2	A	By counting the number of cubes in the figure, you can find that the volume is 30 units^3. There are 3 rows of 8 cubes each in the front part of the figure (3 x 8 = 24) and 3 rows of 2 cubes each at the back of the figure (3 x 2 = 6). Therefore, 24 + 6 = 30.

Question No.	Answer	Detailed Explanation
3	D	By counting the number of cubes in the figure, you can find that the volume is 24 units^3. There bottom layer is 4 by 3 units, so it has a volume of 12 units^3. Each of the top 2 layers is 2 by 3 units, so they each have a volume of 6 units^3. Therefore, 12 + 6 + 6 = 24 units^3.
4	B	The base of the tower measures 4 x 3 x 5, which gives it a volume of 60 cm^3. Since each layer is 4 x 3, it has a volume of 12 cm^3. In order to reach 84 cm^3, Trevor must add 2 more layers (2 x 12 = 24 and 24 + 60 = 84).
5	D	Kerry's figure has a volume of 16 cubic units (you can see 8 cubes at the front of the figure and there are another 8 behind). Trevor's figure has a volume of 24 cubic units (There bottom layer is 4 by 3 units, so it has a volume of 12 units^3. Each of the top 2 layers is 2 by 3 units, so they each have a volume of 6 units^3. Therefore, 12 + 6 + 6 = 24 units^3.) Together, their tower would be 40 cubic units (16 + 24 = 40).

Multiply to Find Volume (5.MD.C.5.A)

Question No.	Answer	Detailed Explanation
1	D	To find the volume of a rectangular prism, multiply the length x width x height (5 x 5 x 8 = 200).
2	C	To find the volume of a rectangular prism, multiply the length x width x height (6 x 3 x 4 = 72).
3	A	Since the volume (66 ft^3) must equal length x width x height, then 2 x 3 x 11 = 66.
4	B	Since the volume (14 in^3) must equal length x width x height, then 7 x 1 x 2 = 14.
5	D	To find the volume of a rectangular prism, multiply the length x width x height (7 x 3 x 2 = 42).

Real World Problems with Volume (5.MD.C.5.B)

Question No.	Answer	Detailed Explanation
1	B	A box that holds 54 1 ft cubes has a volume of 54 ft^3. To find the box with this volume, multiply the length x width x height (3 x 2 x 9 = 54).
2	A	The number of 1 inch cubes it can hold is equal to its volume in inches^3. To find the volume of the container, multiply the area of the base (30) by its height (5). 30 x 5 = 150.

Question No.	Answer	Detailed Explanation
3	C	The volume of the container (300 cm^3) is equal to the area of the base (25) times its height. Therefore, 300 = 25 x 12.
4	C	To find the volume of a rectangular prism, multiply the length x width x height (12 x 5 x 5 = 300). This is greater than the other three jewelry boxes: 10 x 7 x 4 = 280 8 x 5 x 5 = 200 14 x 2 x 10 = 280
5	D	The volume of the cabinet (20 ft^3) will be equal to the area of the base (4 x 1 = 4) times its height. Therefore, the height should be 5 ft (20 = 4 x 5).

Adding Volumes (5.MD.C.5.C)

Question No.	Answer	Detailed Explanation
1	A	To find the total volume, add the volume of the refrigerator (2 x 3 x 4 = 24) to the volume of the freezer (2 x 3 x 2 = 12). 24 + 12 = 36
2	B	To find the volume of one cooler, multiply the length x width x height (30 x 10 x 15 = 4,500). Since Matthew has two coolers, add the two identical volumes: 4,500 + 4,500 = 9,000
3	D	To find the total volume (the number of cubes she can pack), add the volume of the first crate (4 x 4 x 4 = 64) to the volume of the second crate (18 x 2 x 1 = 36). 64 + 36 = 100
4	C	Since V = b x h, the bookcases must be 6 ft tall together (48 = 8 x 6). Therefore, each bookcase must be 3 ft tall (3 + 3 = 6).
5	A	To find the total volume, add the volume of the first box (6 x 3 x 10 = 180) to the volume of the second box (4 x 2 x 2 = 16). 180 + 16 = 196

Name: ________________________ Date: ________________

Geometry

Coordinate Geometry (5.G.A.1)

1. **Assume Point D was added to the grid so that Shape ABCD was a rectangle. Which of these could be the ordered pair for Point D?**

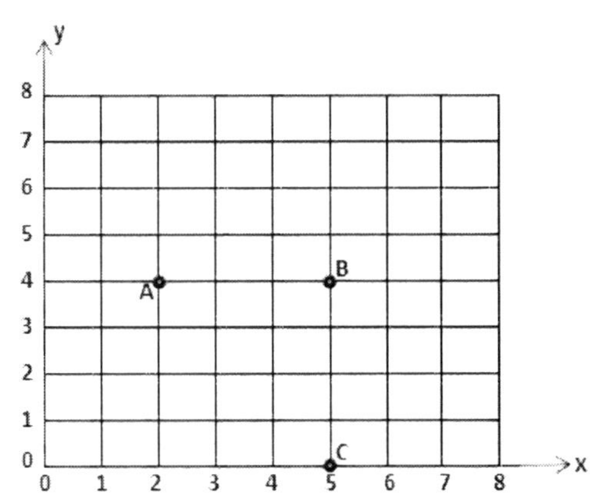

Ⓐ **(0, 0)**
Ⓑ **(0, 2)**
Ⓒ **(2, 0)**
Ⓓ **(2, 2)**

2. **Assume Segments AB and BC were drawn. Compare the lengths of the two segments.**

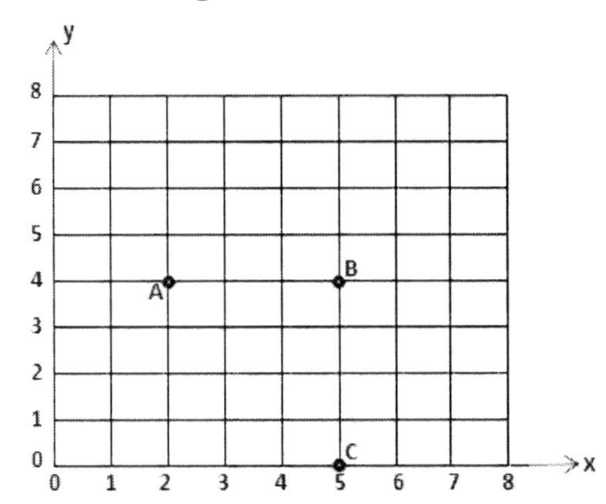

Ⓐ **Segment AB is longer than Segment BC.**
Ⓑ **Segment BC is longer than Segment AB.**
Ⓒ **Segments AB and BC have the same length.**
Ⓓ **It cannot be determined from this information.**

3.

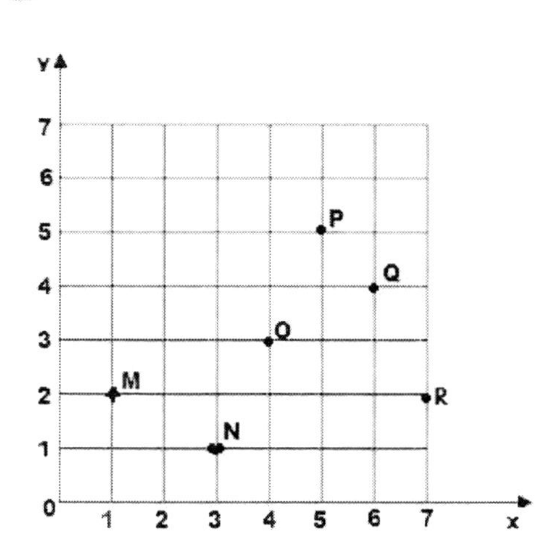

Where is Point R located?

Ⓐ **(2, 7)**
Ⓑ **(7, 2)**
Ⓒ **(6, 4)**
Ⓓ **(4, 6)**

4. **Which point is located at (4, 3)?**

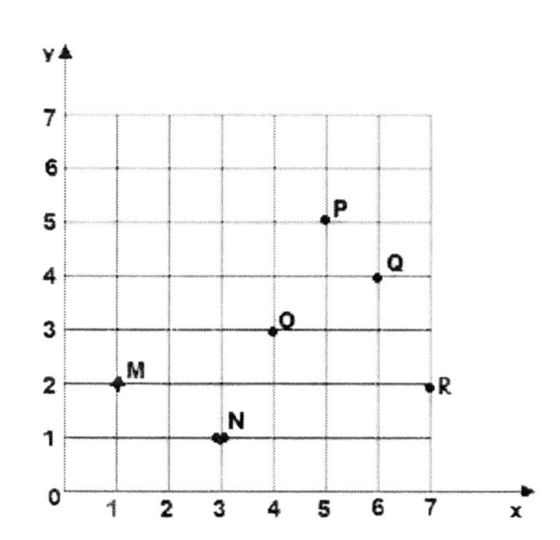

Ⓐ **Point N**
Ⓑ **Point P**
Ⓒ **Point Q**
Ⓓ **Point O**

5. **Locate Point P on the coordinate grid. Which of the following ordered pairs names its position?**

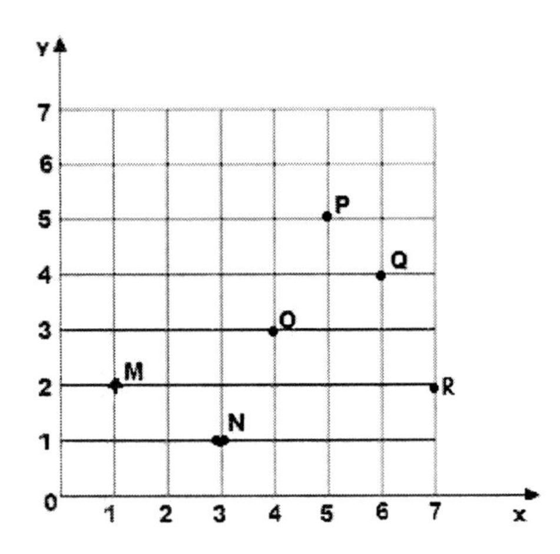

Ⓐ (5,5)
Ⓑ (3, 1)
Ⓒ (1, 2)
Ⓓ (7, 2)

6. **The graph below represents the values listed in the accompanying table, and their linear relationship. Use the graph and the table to respond to the following: What is the value of c (in the table)?**

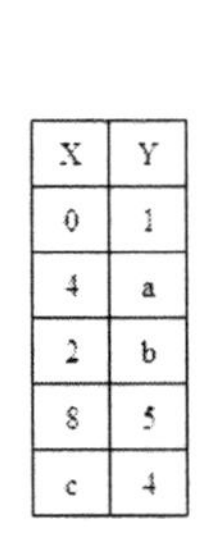

X	Y
0	1
4	a
2	b
8	5
c	4

Ⓐ c = 9
Ⓑ c = 7
Ⓒ c = 6
Ⓓ c = 5

7. The graph below represents the values listed in the accompanying table, and their linear relationship. Use the graph and the table to respond to the following: What is the value of b (in the table)?

X	Y
0	1
4	a
2	b
8	5
c	4

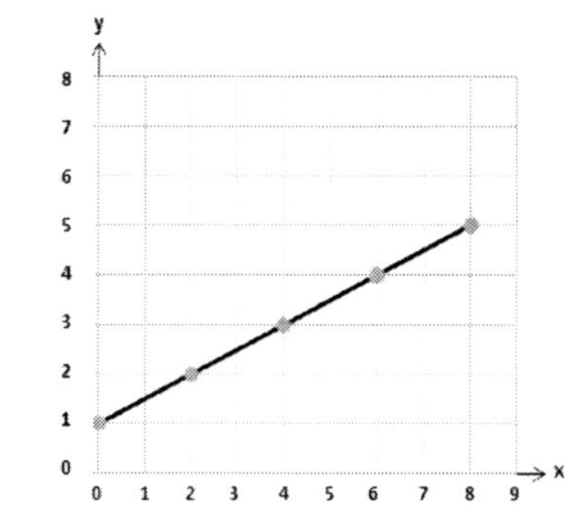

Ⓐ b = 1
Ⓑ b = 2
Ⓒ b = 3
Ⓓ b = 6

8. Which of the following graphs best represents the values in this table?

x	y
1	1
2	2
3	3

Ⓐ

Ⓑ

Ⓒ

Ⓓ

9. On a coordinate grid, which of these points would be closest to the origin?

Ⓐ (2, 1)
Ⓑ (2, 7)
Ⓒ (1, 5)
Ⓓ (0, 4)

10. If these four coordinate pairs were plotted to form a diamond, which point would be the top of the diamond?

Ⓐ (2, 4)
Ⓑ (5, 8)
Ⓒ (8, 4)
Ⓓ (5, 0)

11. Points A (3, 2), B (6, 2), C (6, 6) and D (3, 7) are plotted in a coordinate grid. What type of polygon is ABCD?

Ⓐ a rectangle
Ⓑ a rhombus
Ⓒ a parallelogram
Ⓓ a trapezoid

12. The graph below represents the values listed in the accompanying table, and their linear relationship. Use the graph and the table to respond to the following: Which of the following rules describes the behavior of this function?

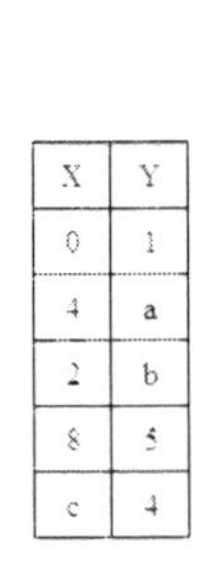

X	Y
0	1
4	a
2	b
8	5
c	4

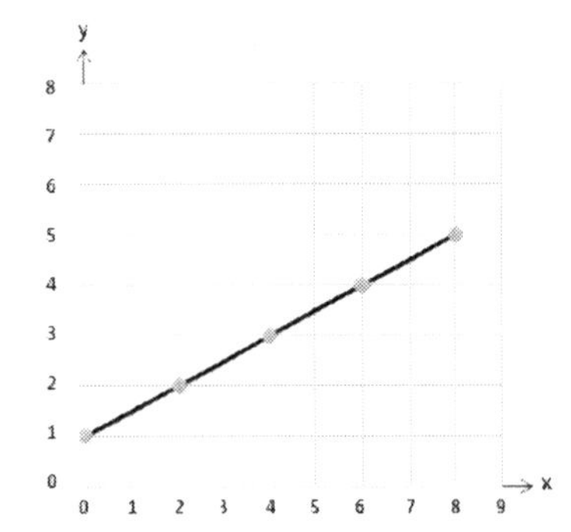

Ⓐ y is equal to one more than x
Ⓑ y is equal to twice x
Ⓒ y is equal to one more than half of x
Ⓓ y is equal to one more than twice x

13.

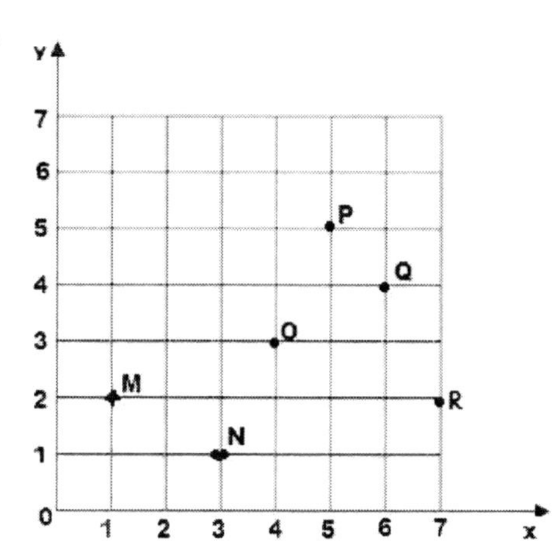

Assume points O, P, and Q form three vertices of a parallelogram. Which ordered pair could be the fourth vertex of that parallelogram?

Ⓐ (3, 3)
Ⓑ (4, 4)
Ⓒ (3, 5)
Ⓓ (3, 4)

14. Which of the following graphs best represents the values in this table?

x	y
3	1
3	2
3	3

Ⓐ

Ⓑ

Ⓒ

Ⓓ

Name: ______________________ Date: ______________

15. If the functions $x = y + 1$ and $y = 3$ were plotted on a coordinate grid, which of these would be true?

Ⓐ **They would intersect once.**
Ⓑ **They would form parallel lines.**
Ⓒ **They would intersect more than once.**
Ⓓ **They would form perpendicular lines.**

Name: ______________________ Date: ______________

Real World Graphing Problems (5.G.A.2)

1. According to the map, what is the location of the weather station (⚡)?

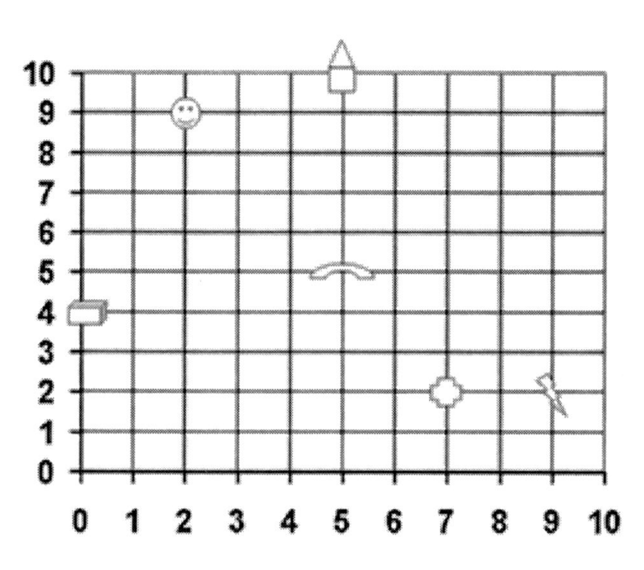

Ⓐ (3,9)
Ⓑ (8,2)
Ⓒ (9,2)
Ⓓ (2,9)

2. According to the map, what is the location of the warehouse (▭)?

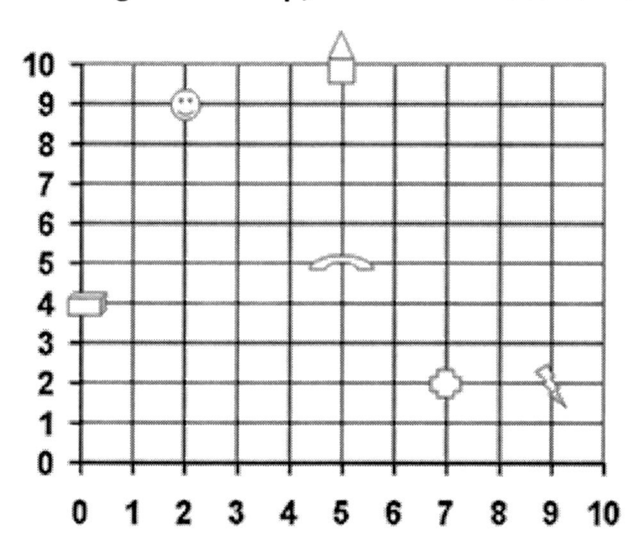

Ⓐ (x = 4)
Ⓑ (0,4)
Ⓒ (y = 4)
Ⓓ (4,0)

Name: ______________________ Date: ______________

3. According to the map, which is located at (7,2)?

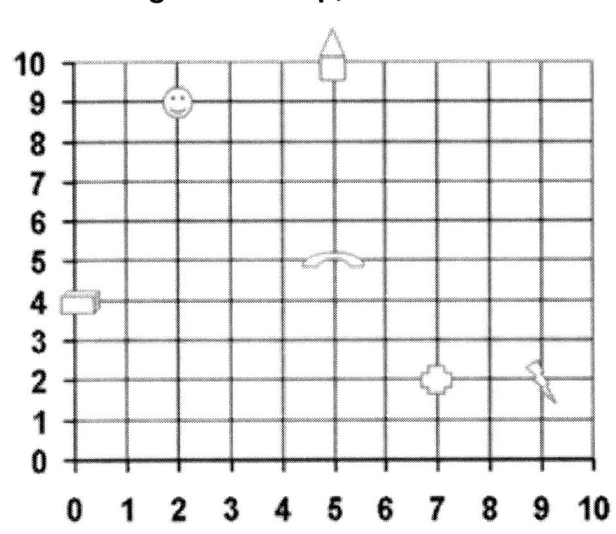

Ⓐ The hospital

Ⓑ The bridge

Ⓒ The playground

Ⓓ The house

4. According to the map, which is located at (5,5)?

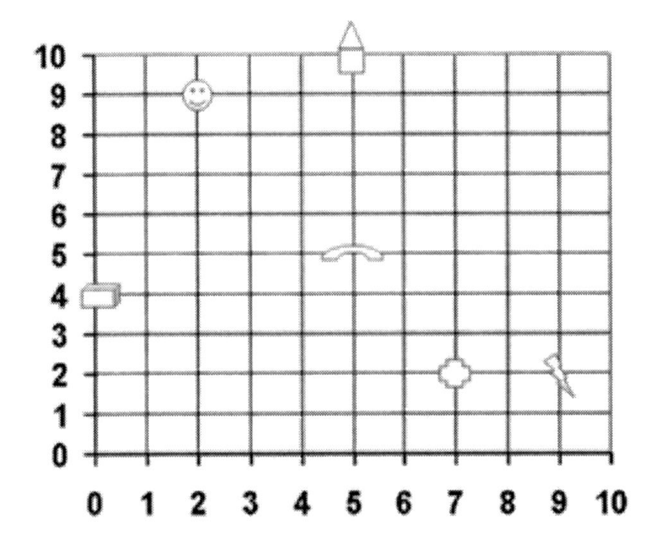

Ⓐ The hospital

Ⓑ The bridge

Ⓒ The playground

Ⓓ The house

5. According to the map, what is the distance from the bridge ( ) to the house ( )?

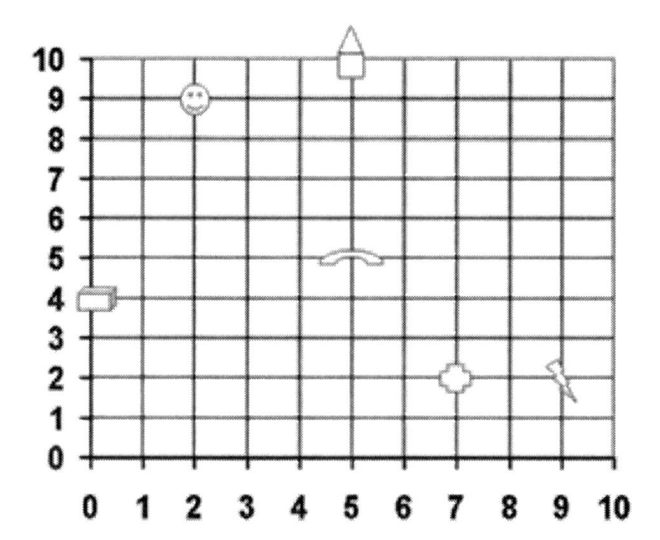

Ⓐ **4 units**
Ⓑ **3 units**
Ⓒ **0 units**
Ⓓ **5 units**

Name: ______________________ Date: ______________

6. Which set of directions would lead a person from the playground (☺) to the hospital (✚)?

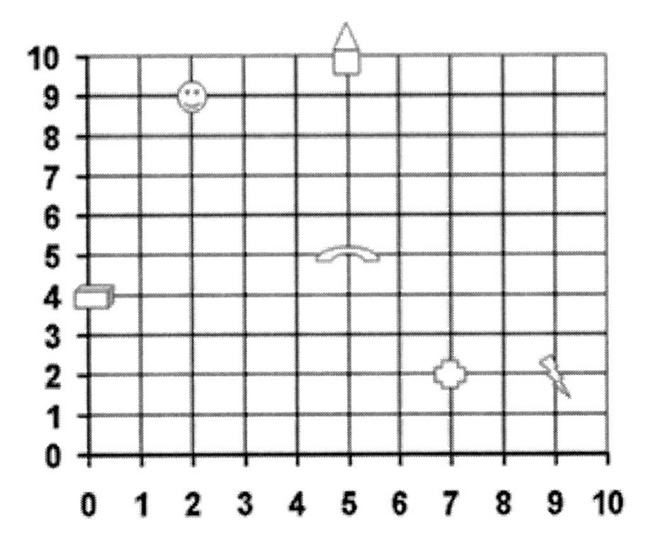

Ⓐ Walk 5 units along the y-axis and 7 units along the x-axis.
Ⓑ Walk 7 units along the y-axis and 2 units along the x-axis.
Ⓒ Walk 2 units along the y-axis and 7 units along the x-axis.
Ⓓ Walk 7 units along the y-axis and 5 units along the x-axis.

7. Which set of directions would lead a person from the weather station (⚡) to the bridge (⌒)?

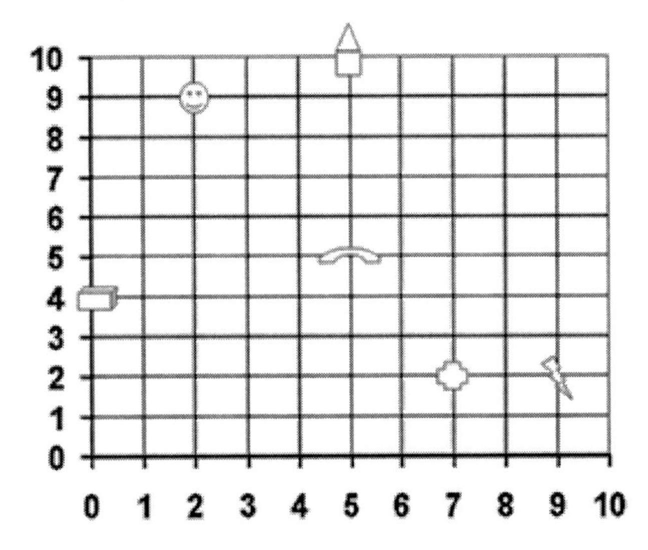

Ⓐ Walk 5 units along the x-axis and 5 units along the y-axis.
Ⓑ Walk 2 units along the x-axis and 0 units along the y-axis.
Ⓒ Walk 4 units along the x-axis and 3 units along the y-axis.
Ⓓ Walk 3 units along the x-axis and 4 units along the y-axis.

8. **Where should the town locate a new lumber mill so it is as close as possible to both the warehouse ( ) and the hospital ( )?**

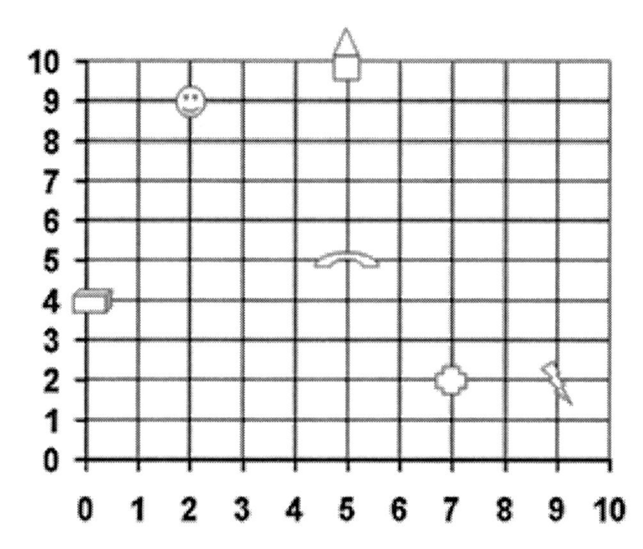

Ⓐ (7,0)
Ⓑ (5,3)
Ⓒ (1,7)
Ⓓ (5,7)

9. **According to the map, what is the location of the zebras ( )?**

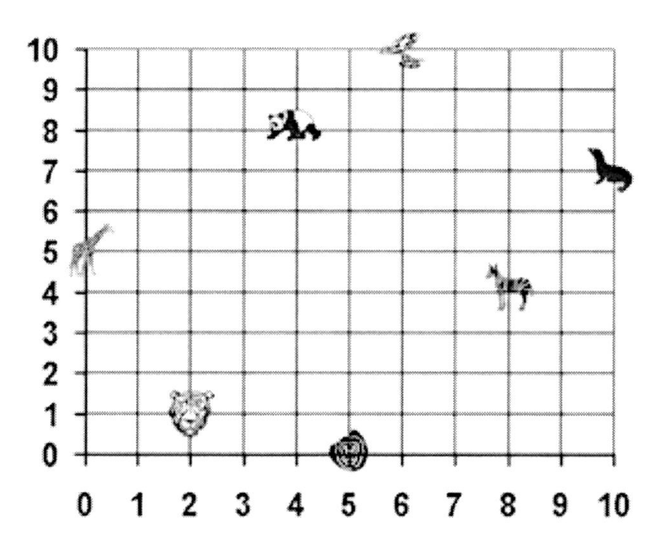

Ⓐ (8,4)
Ⓑ (8,0)
Ⓒ (4,8)
Ⓓ (4,4)

Name: ______________________ Date: ______________

10. According to the map, what is the location of the giraffes ()?

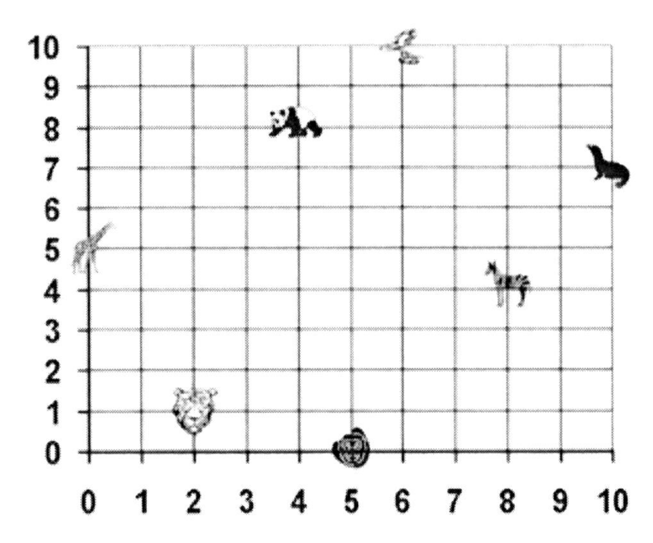

Ⓐ (y = 5)
Ⓑ (0,5)
Ⓒ (x = 5)
Ⓓ (5,0)

11. According to the map, which is located at (10,7)?

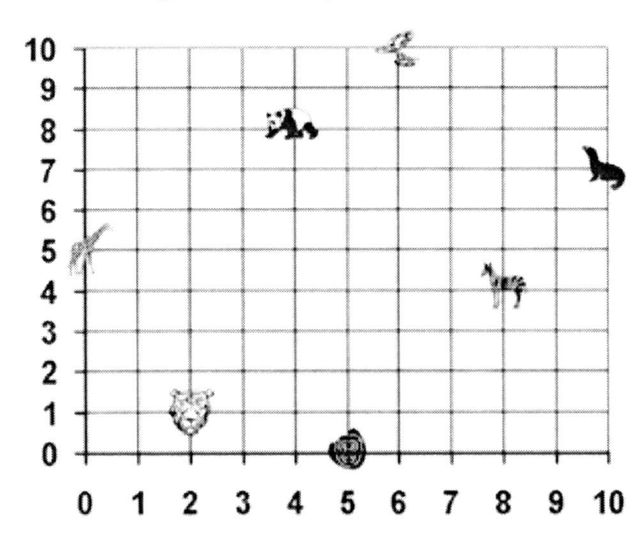

Ⓐ The tigers

Ⓑ The pandas

Ⓒ The snakes

Ⓓ The seals

12. According to the map, which is located at (6,10)?

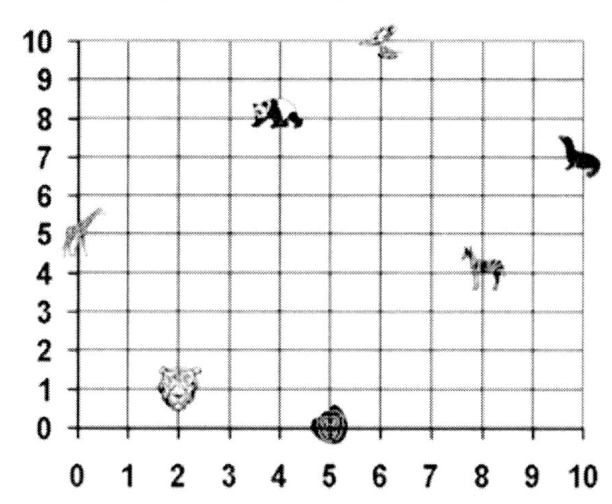

Ⓐ The seals

Ⓑ The monkeys

Ⓒ The snakes

Ⓓ The zebras

13. According to the map, what is the distance from the pandas ( ) to the monkeys ( )?

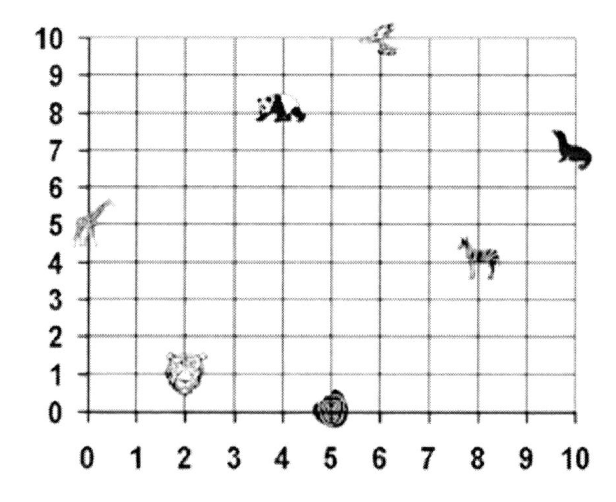

Ⓐ **9 units**
Ⓑ **8 units**
Ⓒ **1 unit**
Ⓓ **6 units**

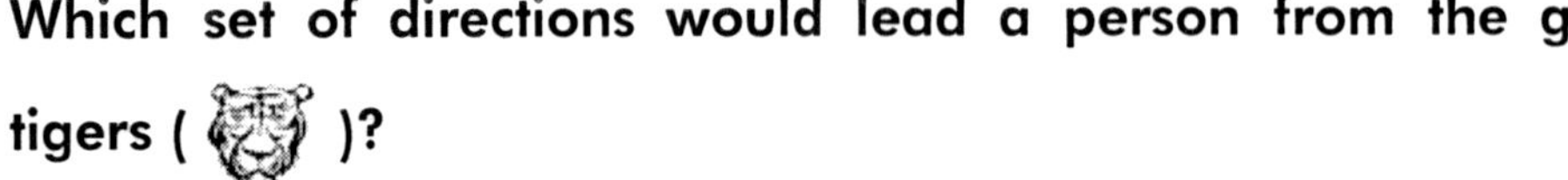

14. Which set of directions would lead a person from the giraffes ( ) to the tigers ( )?

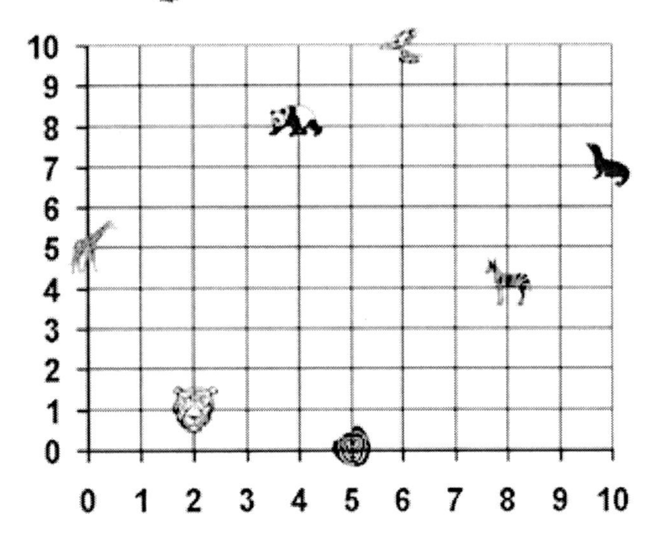

Ⓐ Walk 4 units along the x-axis and 2 units along the y-axis.
Ⓑ Walk 1 unit along the x-axis and 2 units along the y-axis.
Ⓒ Walk 2 units along the x-axis and 4 units along the y-axis.
Ⓓ Walk 4 units along the x-axis and 1 unit along the y-axis.

15. Which would be the best location for the antelopes, so they are as far as possible from the tigers ( )?

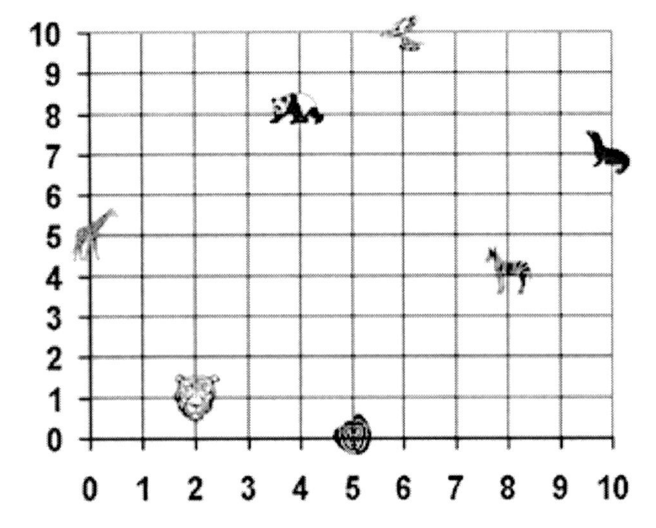

Ⓐ (3,3)
Ⓑ (0,4)
Ⓒ (6,1)
Ⓓ (8,8)

Name: ______________________ Date: ______________

Classify Two-Dimensional Figures into Categories Based on Their Properties

Properties of 2D Shapes (5.G.B.3)

1. **Complete the following.**
   A plane figure is regular only if it has ___________.

   Ⓐ equal sides
   Ⓑ congruent angles
   Ⓒ equal sides and congruent angles
   Ⓓ equal sides, congruent angles, and interior angles that total 180

2. **Complete the following.**
   Two _______ will always be similar.

   Ⓐ circles
   Ⓑ squares
   Ⓒ equilateral triangles
   Ⓓ All of the above

3. **Two interior angles of a triangle measure 30 degrees and 50 degrees. Which type of triangle could it be?**

   Ⓐ a right triangle
   Ⓑ an acute triangle
   Ⓒ an obtuse triangle
   Ⓓ an isosceles triangle

4. **Complete the following.**
   An angle measuring between 0 and 90 degrees is called a(n) _________.

   Ⓐ acute angle
   Ⓑ obtuse angle
   Ⓒ straight angle
   Ⓓ reflex angle

5. **Which of these is not a characteristic of a polygon?**

   Ⓐ a closed shape
   Ⓑ parallel faces
   Ⓒ made of straight lines
   Ⓓ two-dimensional

Name: ______________________ Date: ______________

6. Complete the following.
This isosceles triangle has _________.

Ⓐ one line of symmetry
Ⓑ two congruent angles
Ⓒ two equal sides
Ⓓ all of the above

7. Marcus used tape and drinking straws to build the outline of a two-dimensional shape. He used four straws in all. Exactly three of the straws were of equal length. What might Marcus have built?

Ⓐ a square
Ⓑ a trapezoid
Ⓒ a rectangle
Ⓓ a rhombus

8. How many pairs of parallel sides does a regular octagon have?

Ⓐ 8
Ⓑ 4
Ⓒ 2
Ⓓ 0

9. A rectangle can be described by all of the following terms except __________.

Ⓐ a parallelogram
Ⓑ a polygon
Ⓒ a prism
Ⓓ a quadrilateral

10. Which of the following statements is not true?

Ⓐ An equilateral triangle must have exactly 3 lines of symmetry.
Ⓑ An equilateral triangle will have at least one 60-degree angle.
Ⓒ An equilateral triangle must have rotational symmetry.
Ⓓ All equilateral triangles are congruent.

Name: ______________________ Date: ______________

11. Which of the following shapes is not a polygon?

Ⓐ semicircle
Ⓑ trapezoid
Ⓒ hexagon
Ⓓ decagon

12. What is the name for a polygon that has an acute exterior angle?

Ⓐ convex
Ⓑ concave
Ⓒ complex
Ⓓ simple

13. How many diagonals does a rectangle have?

Ⓐ 1
Ⓑ 2
Ⓒ 3
Ⓓ 4

14. Which shape is a polygon, a quadrilateral, and a rhombus?

Ⓐ an isosceles triangle
Ⓑ a rectangle
Ⓒ a square
Ⓓ a trapezoid

15. Which of the following terms does not describe a trapezoid?

Ⓐ a parallelogram
Ⓑ a polygon
Ⓒ a quadrilateral
Ⓓ a quadrangle

Name: ______________________ Date: ______________

Classifying 2D Shapes (5.G.B.4)

1. **Which shape belongs in the center of the diagram?**

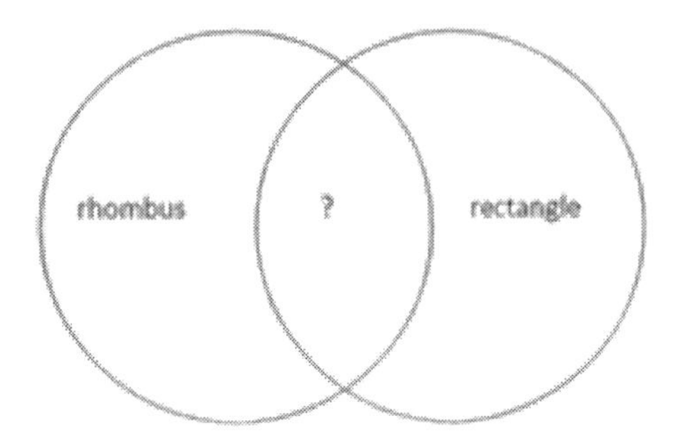

Ⓐ **triangle**
Ⓑ **circle**
Ⓒ **square**
Ⓓ **polygon**

2. **Which shape belongs in section B of the diagram?**

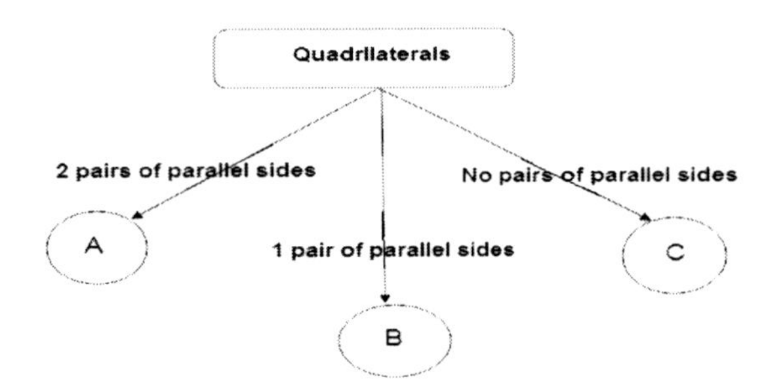

Ⓐ **triangle**
Ⓑ **kite**
Ⓒ **rectangle**
Ⓓ **trapezoid**

3. **Which shape belongs in section C of the diagram?**

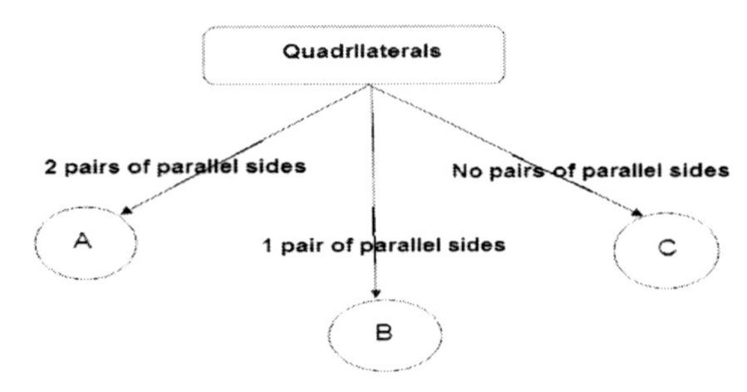

Ⓐ **triangle**
Ⓑ **kite**
Ⓒ **rectangle**
Ⓓ **trapezoid**

4. **Which shape does not belong in section A of the diagram?**

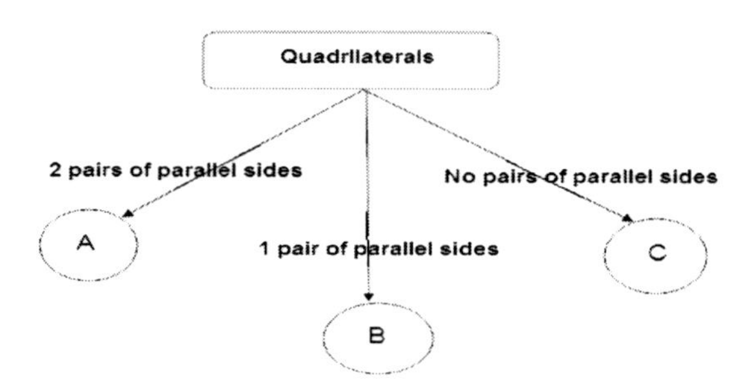

Ⓐ **rectangle**
Ⓑ **diamond**
Ⓒ **rhombus**
Ⓓ **square**

5. **Which shape belongs in section A of the diagram?**

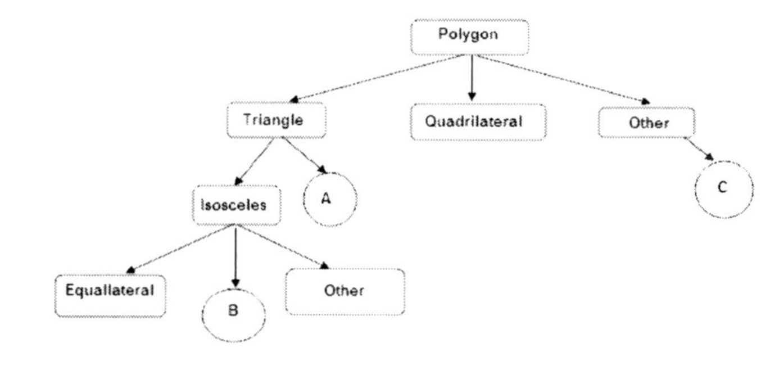

Ⓐ **Scalene**
Ⓑ **Right**
Ⓒ **Acute**
Ⓓ **Symmetrical**

6. **Which shape belongs in section B of the diagram?**

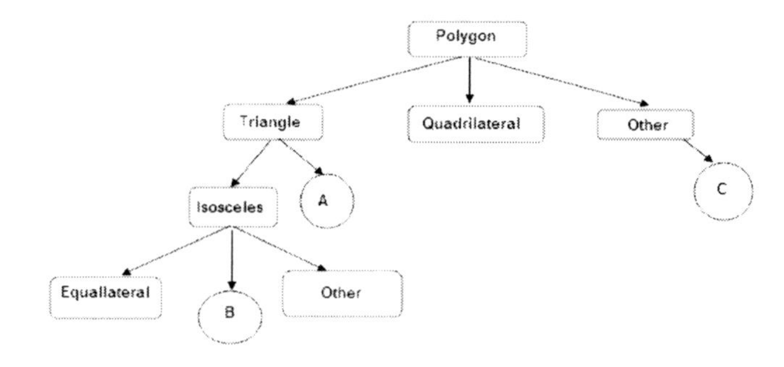

Ⓐ **Scalene**
Ⓑ **Right**
Ⓒ **Acute**
Ⓓ **Symmetrical**

Name: ______________________ Date: ______________

7. **Which shape does not belong in section C of the diagram?**

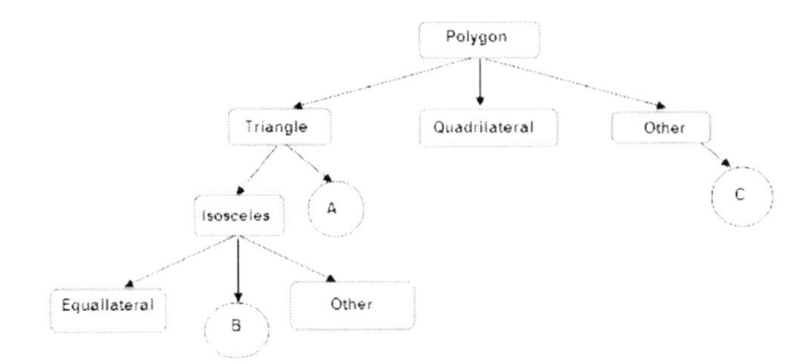

Ⓐ **Octagon**
Ⓑ **Decagon**
Ⓒ **Pentagon**
Ⓓ **Rhombus**

8. **Which statement is true?**

Ⓐ **All triangles are quadrangles.**
Ⓑ **All polygons are hexagons.**
Ⓒ **All parallelograms are polygons.**
Ⓓ **All parallelograms are rectangles.**

9. **Which statement is false?**

Ⓐ **All octagons are parallelograms.**
Ⓑ **All rectangles are quadrangles.**
Ⓒ **All triangles polygons.**
Ⓓ **All quadrilaterals are rectangles.**

10. **Which statement defines a quadrangle?**

Ⓐ **Any rectangle that has 4 sides of equal length**
Ⓑ **Any quadrilateral with 2 pairs of parallel sides**
Ⓒ **Any polygon with 4 angles**
Ⓓ **Any polygon with 4 or more sides**

11. Which shape is a quadrilateral but not a parallelogram?

Ⓐ 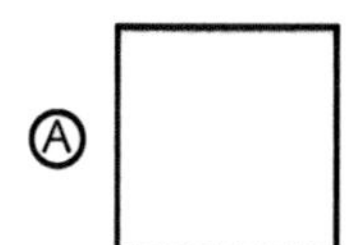

Ⓑ 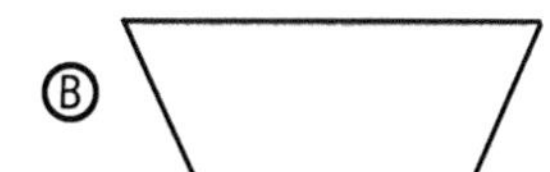

Ⓒ 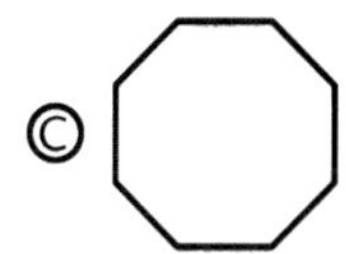

Ⓓ 

12. Which shape is a rhombus but not a rectangle?

Ⓐ 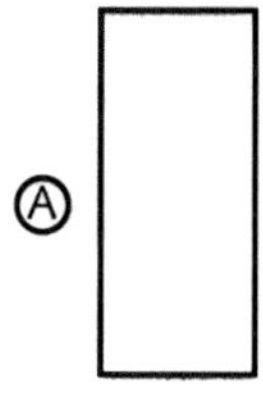

Ⓑ 

Ⓒ 

Ⓓ 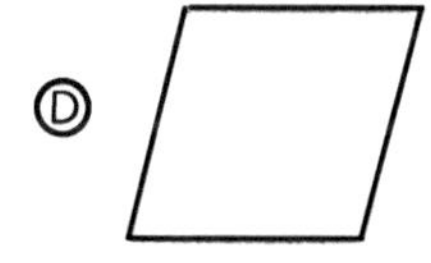

13. In a hierarchy of shapes, how could the category of "pentagon" be split in two?

Ⓐ Polygon and non-polygon
Ⓑ 5 sides and 6 sides
Ⓒ Parallelogram and not parallelogram
Ⓓ Regular and non-regular

14. Which triangle would be classified as equiangular?

Ⓐ 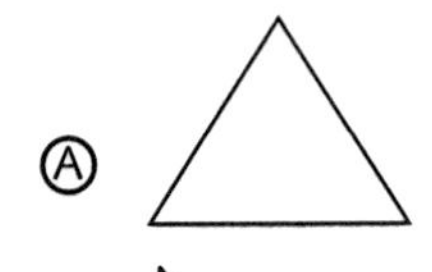

Ⓑ 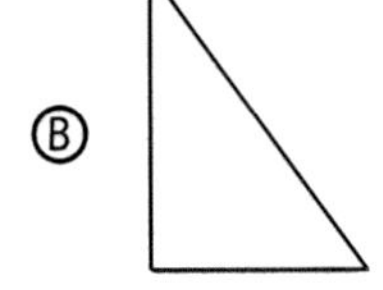

Ⓒ 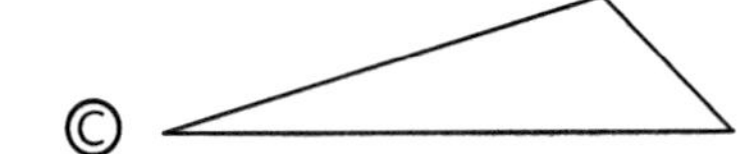

Ⓓ 

15. Which shape would not be classified as regular?

Ⓐ 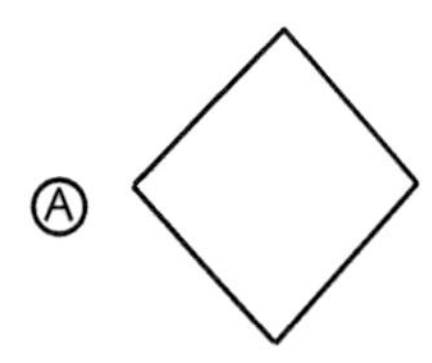

Ⓑ 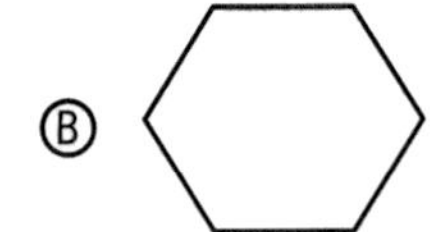

Ⓒ 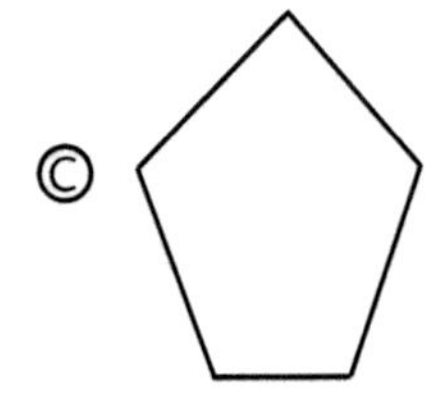

Ⓓ 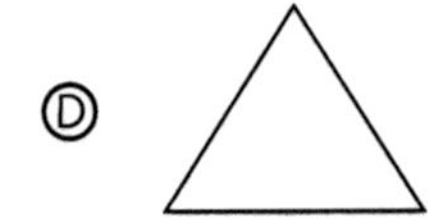

End of Geometry

Geometry

Answer Key & Detailed Explanations

Coordinate Geometry (5.G.A.1)

Question No.	Answer	Detailed Explanation
1	C	A rectangle must have two pairs of parallel sides, so point D must be at 2 on the x-axis (in line with point A) and at 0 on the y-axis (in line with point C).
2	B	Segment AB (from 2 to 5) is 3 units long. Segment BC (from 0 to 4) is 4 units long. Segment BC is longer.
3	B	Using the labels, follow the x-axis as far as point R (7 units) and the y-axis as far as point R (2 units). This makes the coordinate pair (7, 2).
4	D	To find point (4, 3), follow the x-axis horizontally 4 units, then follow the y-axis vertically 3 units. The result is point O.
5	A	Using the labels, follow the x-axis as far as point P (5 units) and the y-axis as far as point P (5 units). This makes the coordinate pair (5, 5).
6	C	The coordinate pair is (c, 4), so follow the line on the graph to where its value for y is 4. Follow that point down to the x-axis to see it is 6.
7	B	The coordinate pair is (2, b), so follow the line on the graph to where its value for x is 2. Follow that point across to the y-axis to see it is 2.
8	B	As the value of x increases, the value of y increases equally. This produces an upward-sloping straight line.
9	A	The origin is point (0, 0). The point closest to this would have the lowest x- and y- values (without being negative numbers).
10	B	The points would create this diamond:

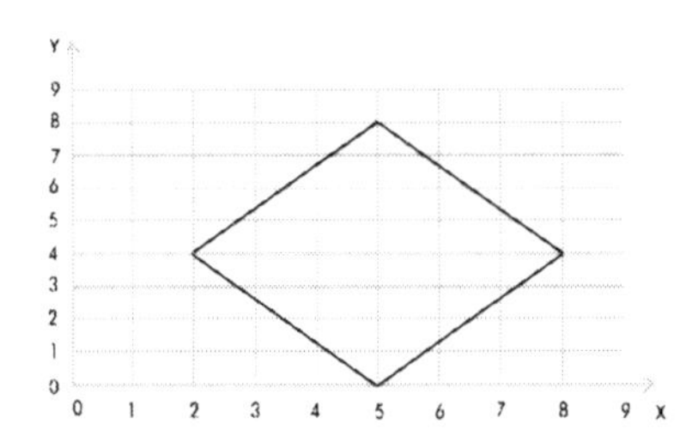

The top of the diamond is point (5, 8).

Question No.	Answer	Detailed Explanation
11	D	When the four points are plotted and connected, they form this polygon: 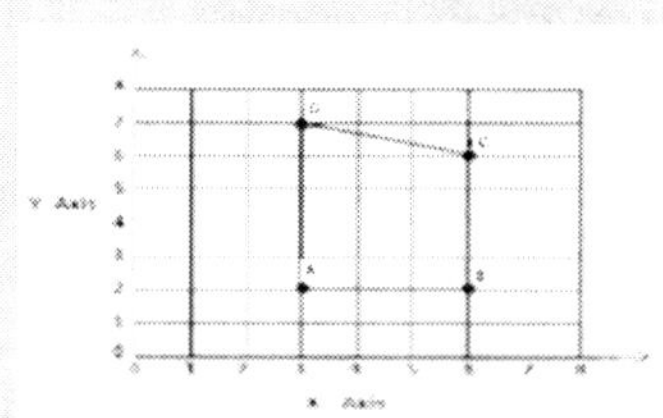 It is a trapezoid.
12	C	The statement y is equal to one more than half of x can be checked with the formula y = x/2 + 1. Plugging in the coordinates (8, 5) gives us 5 = 8/2 + 1, which is true.
13	D	Naming point (3, 4) point L would make segment LO parallel to segment PQ and it would make segment LP parallel to segment OQ.
14	C	In this example, one value for x has three different values for y. This will create a straight vertical line.
15	A	The functions would create these two intersecting lines: 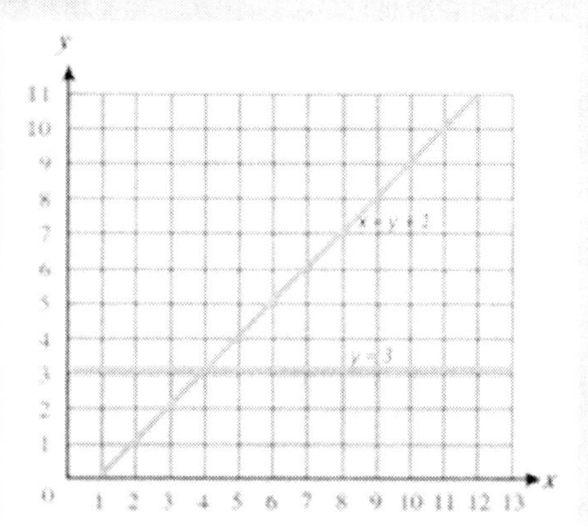

Real World Graphing Problems (5.G.A.2)

Question No.	Answer	Detailed Explanation
1	C	The location of the weather station is at the intersection of 9 on the x-axis and 2 on the y-axis. Therefore, its coordinates are (9,2).
2	B	The location of the warehouse is at the intersection of 0 on the x-axis and 4 on the y-axis. Therefore, its coordinates are (0,4).
3	A	The location of the hospital is at the intersection of 7 on the x-axis and 2 on the y-axis. Therefore, its coordinates are (7,2).

Question No.	Answer	Detailed Explanation
4	B	The location of the bridge is at the intersection of 5 on the x-axis and 5 on the y-axis. Therefore, its coordinates are (5,5).
5	D	Since both the bridge and the house are located at 5 on the x-axis, find their distance along the y-axis. The house is located at 10 on the y-axis and the bridge is located at 5 on the y-axis. The difference is 5 units.
6	D	Starting at the playground (2,9), walking 7 units along the y-axis could bring a person to (2,2). From there, walking 5 units along the x-axis could bring that person to (7,2), to the location of the hospital.
7	C	Starting at the weather station (9,2), walking 4 units along the x-axis could bring a person to (5,2). From there, walking 3 units along the y-axis could bring that person to (5,5), to the location of the bridge.
8	B	This is the only set of coordinates given that is located between the warehouse and the hospital, making it the closest to both locations.
9	A	The location of the zebras is at the intersection of 8 on the x-axis and 4 on the y-axis. Therefore, its coordinates are (8,4).
10	B	The location of the giraffes is at the intersection of 0 on the x-axis and 5 on the y-axis. Therefore, its coordinates are (0,5).
11	D	The location of the seals is at the intersection of 10 on the x-axis and 7 on the y-axis. Therefore, its coordinates are (10,7).
12	C	The location of the snakes is at the intersection of 6 on the x-axis and 10 on the y-axis. Therefore, its coordinates are (6,10).
13	A	The pandas are located at 8 on the y-axis and the monkeys are located at 0 on the y-axis. The difference is 8 units. The pandas are located at 4 on the y-axis and the monkeys are located at 5 on the y-axis. The difference is 1 unit. All together, the distance is 9 units (8 + 1 = 9)

Question No.	Answer	Detailed Explanation
14	C	Starting at the giraffes (0,5), walking 2 units along the x-axis could bring a person to (2,5). From there, walking 4 units along the y-axis could bring that person to (2,1), to the location of the tigers.
15	D	This is the furthest location from the tigers, which are located at (2,1).

Properties of 2D Shapes (5.G.B.3)

Question No.	Answer	Detailed Explanation
1	C	In a regular plane figure, all sides are equal lengths and all angles are congruent. However the angles do not have to total 180.
2	D	Similar shapes do not have to be the same size, but they must be the same shape. All sides must have the same ratio. Since circles, squares, and equilateral triangles always have the same ratio of sides (or dimension, in the case of the circle) they are always similar.
3	C	Remember that the three interior angles of a triangle always equal 180 degrees. If two of the angles equal 80 degrees, the third angle must be 100 degrees. Any triangle with an angle greater than 90 degrees is obtuse.
4	A	An acute angle is one that is less than 90 degrees.
5	B	Since polygons are two-dimensional, they do not have faces.
6	D	By definition, an isosceles triangle has two equal sides and two congruent angles. It also has one line of symmetry (in this model, a vertical line through the center).
7	B	Marcus could not have made a square or a rhombus, since only three of the sides were of equal length. He also could not have made a rectangle, since the number of equal sides on a rectangle is either 2 or 4 (if it is a square). The shape he made must have been an (isosceles) trapezoid.
8	B	All of the opposite sides of a regular octagon are parallel. Since there are 8 sides, there are 4 pairs.
9	C	A prism is a 3-dimensional shape. Even though it can consist of rectangular faces, a rectangle is not a prism.

Question No.	Answer	Detailed Explanation
10	D	In order to be congruent, all equilateral triangles would have to be the same size, which they are not. Equilateral triangles of any size do have 3 lines of symmetry, 3 60-degree angles, and rotational symmetry.
11	A	A polygon is a closed 2-dimensional figure that is made entirely of straight lines. Since a semicircle has a curved side, it is not a polygon.
12	B	A concave polygon has at least one exterior angle that is acute (or one interior angle that is obtuse).
13	B	A diagonal is a straight line that connects one corner of a polygon with another corner (but is not a side). All quadrilaterals have two diagonals.
14	C	While it is a polygon and a quadrilateral, it is the fact that it is a rhombus (4 equal-length sides) that makes this figure a square.
15	A	A trapezoid is a flat, closed figure made of straight lines (a polygon). It has four sides (a quadrilateral) and four angles (a quadrangle). It has only one set of opposite sides parallel, so it is not a parallelogram (two pairs of parallel sides).

Classifying 2D Shapes

Question No.	Answer	Detailed Explanation
1	C	A square is both a rhombus and a rectangle.
2	D	A trapezoid is a quadrilateral with one pair of parallel sides.
3	B	A kite is a quadrilateral with no pairs of parallel sides.
4	B	A diamond is a quadrilateral but it does not have two pairs of parallel sides.
5	A	Scalene is a type of triangle that is not isosceles.
6	B	A right triangle (one with a 90° angle) is a type of isosceles triangle.
7	D	A rhombus has four sides, so it would fall under the heading quadrilateral in the hierarchy.
8	C	A polygon is a closed 2-dimensional figure made up of straight lines. Therefore, all parallelograms are polygons.

Question No.	Answer	Detailed Explanation
9	A	An octagon is any 8-sided polygon. Parallelograms have 4 sides.
10	C	A quadrangle is simply any closed 2-dimensional figure made up of straight lines (a polygon) that has 4 sides or 4 angles. Options A and B fall under the classification of quadrangle, but they do not define it entirely.
11	B	The trapezoid is a quadrilateral because it has four sides, but it is not a parallelogram because it does not have 2 sets of parallel sides.
12	D	This is a rhombus because it is a parallelogram with 4 sides of equal length, but it is not a rectangle because it does not have 4 right angles.
13	D	A pentagon can be either regular (having all sides and angles the same) or non-regular (having 5 sides and angles that differ).
14	A	An equiangular triangle is one in which all three angles are equal (also called an equilateral triangle because the three sides are the same length).
15	C	A regular polygon is one in which all angles are the same and all sides are the same length. This pentagon has sides and angles that differ.

Notes

Notes

INCLUDES Online Workbooks!

About Online Workbooks

- When you buy this book, 1 year access to online workbooks is included
- Access them anytime from a computer with an internet connection
- Adheres to the Common Core State Standards
- Includes progress reports
- Instant feedback and self-paced
- Ability to review incorrect answers
- Parents and Teachers can assist in student's learning by reviewing their areas of difficulty

Course Name: Grade 4 Math Prep

Lesson Name:	Correct	Total	% Score	Incorrect
Introduction				
Diagnostic Test		3	0%	3
Number and Numerical Operations				
Workbook - Number Sense	2	10	20%	8
Workbook - Numerical Operations	2	25	8%	23
Workbook - Estimation	1	3	33%	2
Geometry and measurement				
Workbook - Geometric Properties		6	0%	6
Workbook - Transforming Shapes				
Workbook - Coordinate Geometry	1	3	33%	2
Workbook - Units of Measurement				
Workbook - Measuring Geometric Objects	3	10	30%	7
Patterns and algebra				
Workbook - Patterns	7	10	70%	3
Workbook - Functions and relationships				

LESSON NAME: Workbook - Geometric Properties

Elapsed Time: 01:10

Question No. 2

What type of motion is being modeled here?

Select right answer

- a translation
- a rotation 90° clockwise
- a rotation 90° counter-clockwise
- a reflection

Previous question | Next question

Report Name: Missed Questions

Student Name: Lisa Colbright
Cours Name: Grade 4 Math Prep
Lesson Name: Diagnostic Test

The faces on a number cube are labeled with the numbers 1 through 6. What is the probability of rolling a number greater than 4?

Answer Explanation

(C) On a standard number cube, there are six possible outcomes. Of those outcomes, 2 of them are greater than 4. Thus, the probability of rolling a number greater than 4 is "2 out of 6" or 2/6.

A)		1/6
B)		1/3
C)	Correct Answer	2/6
D)		3/6

CPSIA information can be obtained
at www.ICGtesting.com
Printed in the USA
LVOW09s2320051216
515975LV00023B/445/P

9 781940 48445